# The Individual's Guide for Understanding and Surviving Terrorism

I0115403

U.S. Marine Corps

Fredonia Books
Amsterdam, The Netherlands

The Individual's Guide for Understanding and
Surviving Terrorism

by
U. S. Marine Corps

ISBN: 1-4101-0021-9

Reprinted from the original edition

Fredonia Books
Amsterdam, The Netherlands
http://www.fredoniabooks.com

ARTM NT F T NA
ead uarters United States Marine Corps
ashington, C 20 0-1

1 September 2001

## FOREWORD

Marine Corps Reference ublication MCR -02 , *The Individual's Guide for Understanding and Surviving Terrorism*, provides guidance to individual Marines private through general officer and their dependents on terrorism and its effects.

This reference publication gives an overview of terrorism, e plains antiterrorism individual protective measures, and what to do if taken hostage.

MCR -02 supersedes Fleet Marine Force Reference ublication FMFR -14A, *The Individual's Guide for Understanding and Surviving Terrorism*, dated 1 ctober 19 9, and FMFR - , *Vehicle Bomb Search*, dated 0 April 1990.

Reviewed and approved this date.

B IR CTI N F T C MMAN ANT F T MARIN C R S

AR AN N, R.
ieutenant General, U.S. Marine Corps
Commanding General
Marine Corps Combat evelopment Command

ISTRIBUTI N: 144 0000 1 00

# The Individual's Guide for Understanding and Surviving Terrorism

## Table of Contents

### Chapter 1. Overview

### Chapter 2. Protection Through Awareness

# Chapter 3. Hostage Survival

## Appendices

# Chapter 1
# Overview

## Terrorism Defined

oint ublication 1-02, *Department of Defense Dictionary of Military and Associated Terms*, defines terrorism as the calculated use of unlawful violence or threat of unlawful violence to inculcate fear intended to coerce or to intimidate governments or societies in the pursuit of goals that are generally political, religious, or ideological. ithin this definition are three key elements violence, fear, and intimidation. ach element produces terror in its victims. The policy of the United States is summari ed as follows:

- All terrorist acts are criminal and intolerable, whatever their motivation, and should be condemned.

- The U.S. will support all lawful measures to prevent terrorism and bring those responsible to ustice.

- No concessions will be made to terrorist e tortion, because to do so only invites more terrorist action.

- hen Americans are abducted overseas, the U.S. will look to the host government to e ercise its responsibility to protect all persons within its territories, to include achieving the safe release of hostages.

- The U.S. will maintain close and continuous contact with the host government during the incident and will continue to develop international cooperation to combat terrorism.

## Strategy

Terrorism is a criminal act that influences an audience beyond the immediate victim. The strategy of terrorists is to commit acts of violence that draw the attention of the local populace, the government, and the world to their cause. Terrorists plan their attack to obtain the greatest publicity, choosing targets that symboli e what they oppose. The effectiveness of the terrorist act lies not in the act itself, but in the public's or government's reaction to the act.

For e ample, at the 19 2 Munich  lympics, the Black September  rgani ation killed 11 Israelis. The Israelis were the immediate victims. But the true target was the estimated 1 billion people watching the televised event. The Black September  rgani ation used the high visibility of the  lympics to publici e its views on the plight of the  alestinian refugees.

Similarly, in  ctober 19 , Middle  astern terrorists bombed the Marine Battalion  anding Team  ead uarters at Beirut International Airport. Their immediate victims were the 241 U.S. military personnel who were killed and over 100 others who were wounded. Their true target was the American people and the U.S. Congress. Their one act of violence influenced the United States' decision to withdraw the Marines from Beirut and was therefore considered a terrorist success.

 n 11 September 2001, terrorists sky acked U.S. commercial planes and crashed two planes into the  orld Trade Center in New  ork City and one into the  entagon in  ashington .C. The terrorist attacks inflicted serious loss of life by destroying the  orld Trade Center towers and part of the  entagon building. They were designed to strike a blow at the American will and its economic and military structure. Although the attacks succeeded in hitting their targets, they galvani ed the will of the American public to take political, financial, and military actions to combat terrorism.

## Perspectives

There are three perspectives of terrorism: the terrorist's, the victim's, and the general public's. The phrase one man's terrorist is another man's freedom fighter is a view terrorists themselves would accept. Terrorists do not see themselves as evil. They believe they are legitimate combatants, fighting for what they believe in, by whatever means possible. A victim of a terrorist act sees the terrorist as a criminal with no regard for human life. The general public's view is the most unstable. The terrorists take great pains to foster a Robin ood image in hope of swaying the general public's point of view toward their cause.

## Today's Threat

Many areas of the world are e periencing great political, economic, and social unrest. The reasons for this unrest can be seen in conflicts with neighboring states, internal strife, dissatisfaction with governments in power, unconstrained population growth, declining resources, and ethnic and religious hatreds. This unrest has spawned numerous groups that lack the means to have their grievances solved by their own governments through the normal political processes. Sometimes these groups resort to terrorism to achieve their aims. Generally, these aims stem from political ideology, nationalism, religion or special interests.

ver the past 20 years, terrorists have committed e tremely violent acts for alleged political or religious reasons. olitical ideology ranges from the far left to the far right. For e ample, the far left can consist of groups such as Mar ists and eninists who propose a revolution of workers led by a revolutionary elite. n the far right, are dictatorships that typically believe in a merging of state and business leadership.

Nationalism is the devotion to the interests or culture of a group of people or a nation. Typically, nationalists share a common ethnic background and wish to establish or regain a homeland.

Religious e tremists often re ect the authority of secular governments and view legal systems that are not based on their religious beliefs as illegitimate. They often view moderni ation efforts as corrupting influences on traditional culture.

Special interest groups include people on the radical fringe of many legitimate causes such as antiabortion views, animal rights, radical environmentalism. These groups believe that violence is morally ustifiable to achieve their goals.

## Types of Terrorist Incidents

The following figure provides a spectrum of terrorist incidents, from the most common to the least common.

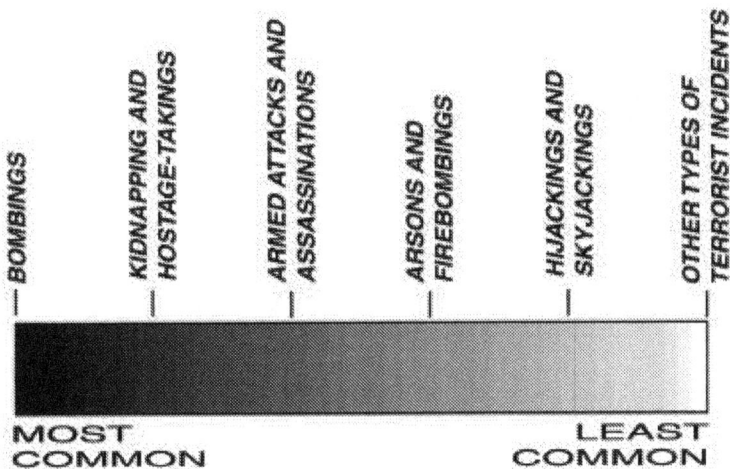

MOST COMMON — LEAST COMMON

BOMBINGS · KIDNAPPING AND HOSTAGE-TAKINGS · ARMED ATTACKS AND ASSASSINATIONS · ARSONS AND FIREBOMBINGS · HIJACKINGS AND SKYJACKINGS · OTHER TYPES OF TERRORIST INCIDENTS

## Bombings

Bombings are the most common type of terrorist act. Typically, improvised e plosive devices are ine pensive and easy to make. Modern devices are smaller and harder to detect. They contain very destructive capabilities  for e ample, on 0 August 199 , two American embassies in Africa were bombed. The bombings claimed the lives of over 200 people, including 12 innocent American citi ens, and in ured over ,000 civilians.

Terrorists can also use materials readily available to the average consumer to construct a bomb. For e ample, on 19 April 199 , the Murrah Federal Building in klahoma City, , was bombed, and 1  people were killed by an improvised e plosive device.

In the case of the 11 September 2001 attacks, terrorists sky acked commercial planes, laden with full fuel tanks, to fly them into buildings as guided missiles or massive flying bombs. This awakened the world to a new level of terrorist bombing efforts.

## Kidnappings and Hostage-Takings

Terrorists use kidnappings and hostage-takings to establish a bargaining position and elicit publicity. Although kidnapping is one of the most difficult acts for a terrorist group to accomplish, if it is successful, it can gain terrorists money, release of ailed comrades, and publicity for an e tended period of time.

ostage-taking involves the sei ure of a facility or location and the taking of hostages. Unlike a kidnapping, hostage-taking provokes a confrontation with authorities. It forces authorities to either make dramatic decisions or to comply with the terrorist's demands. ostage-taking is overt and designed to attract and hold media attention. The terrorists' intended target is the audience affected by the hostage's confinement, not the hostage himself.

## Armed Attacks and Assassinations

Armed attacks include raids and ambushes. Assassinations are the killing of a selected victim, usually by bombings or small arms. rive-by shootings are a common techni ue employed by loosely organi ed terrorist groups. istorically, terrorists have assassinated specific individuals for psychological effect.

## Arsons and Firebombings

Incendiary devices are cheap and easy to hide. Arson and firebombings are easily conducted by terrorist groups that may not be

as well-organi ed, e uipped or trained as a ma or terrorist organi-
ation. Arsons or firebombings against utilities, hotels, government buildings or industrial centers are common tactics used by
terrorists to portray an image that the ruling government is inca-
pable of maintaining order.

## Hijackings and Skyjackings

i acking is the sei ure by force of a surface vehicle, its passengers, and/or its cargo. Sky acking is the taking of an aircraft, which
creates a mobile, hostage barricade situation. It provides terrorists
with hostages from many nations and draws heavy media attention.
Sky acking also provides mobility for the terrorists to relocate the
aircraft to a country that supports their cause and provides them
with a human shield, making retaliation difficult.

n 11 September 2001, commercial airplanes were sky acked but
only to gain control of the aircraft. The terrorists' intent was not to
create a hostage barricade situation, but to ensure the passengers
did not interfere with their desire to crash the aircraft into their
intended targets.

## Other Types of Terrorist Incidents

In addition to the acts of violence discussed, numerous other
types of violence e ist under the framework of terrorism. Terrorist groups conduct maimings against their own people as a form
of punishment for security violations, defections or informing.
Terrorist organi ations also conduct robberies and e tortion when
they need to finance their acts and are without sponsorship from
sympathetic nations.

Cyberterrorism is a new, increasing form of terrorism that targets computer networks. Cyberterrorism allows terrorists to conduct their operations with little or no risk to themselves. It also provides terrorists an opportunity to disrupt or destroy networks and computers. The result is interruption of key government or business-related activities. Although this type of terrorism lacks a high profile compared to other types of terrorist attacks, its impact is ust as destructive.

istorically, terrorist attacks using nuclear, biological, and chemical NBC weapons have been rare. ue to the e tremely high number of casualties that NBC weapons produce, NBC weapons are also referred to as weapons of mass destruction M . A number of nations are involved in arms races with neighboring countries because they view the development of M as a key deterrent of attack by hostile neighbors. The increased development of M also increases the potential for terrorist groups to gain access to M . It is believed that in the future terrorists will have greater access to M because unstable nations or states may fail to safeguard their stockpiles of M from accidental losses, illicit sales or outright theft or sei ure.

etermined terrorist groups can also gain access to M through covert independent research efforts or by hiring technically skilled professionals to construct them. Although an e plosive nuclear device is believed beyond the scope of most terrorist groups, chemical, biological or radiological dispersion weapons that use nuclear contaminants are not.

An e ample of a terrorist group gaining access to M was tragically evident on 20 March 199 . A apanese religious cult, Aum Shinrikyo or Supreme Truth, chemically attacked apanese citi ens in the Tokyo subway system. Use of the nerve agent Sarin resulted in 12 deaths and , 00 hospitali ations.

In ctober of 2001, several letters were mailed to selected U.S. Government and media individuals. Those letters contained the biological agent anthra . In large amounts or even small amounts widely distributed, such biological agents can be a M . Fear of these biological agents can create as much terrorist value as their actual employment.

## Intentions of Terrorist Groups

- roduce widespread fear.

- btain worldwide, national or local recognition for their cause by attracting the attention of the media.

- arass, weaken, or embarrass government security forces so that the government overreacts and appears repressive.

- Steal or e tort money and e uipment, especially weapons and ammunition.

- estroy facilities or disrupt lines of communication in order to create doubt that the government can provide for and protect its citi ens.

- iscourage foreign investments, tourism or assistance programs that can affect the target country's economy and support of the government in power.

- Influence government decisions, legislation or other critical decisions.

- Free prisoners.

- Satisfy vengeance.

- Turn the tide in a guerrilla war by forcing government security forces to concentrate their efforts in urban areas. This allows the terrorist groups to establish themselves among the local populace in rural areas.

## General Characteristics of Terrorist Groups

- Seek to intimidate by promoting fear.
- Some have advanced weaponry e.g., tanks but generally they are militarily weaker than the governments they fight.
- mploy unconventional warfare tactics training can include physical and mental preparation, weapons and e plosives, political and religious indoctrination, combat tactics, intelligence, psychological warfare, survival, and communications.
- o not e uate tactical success with mission success. A specific terrorist act may not achieve its desired results, but a terrorist may still view the act as successful if it publici es the cause.
- Usually urban-based and highly mobile. If urban-based, terrorists have access to mass transportation e.g., airplanes, ships, railroads, and subways . Terrorist groups with international contacts may also have access to forged passports and safehavens in other countries.
- Generally organi e and operate clandestinely in cells of three to five members. A cell may only have contact with another cell or the ne t higher command level. Therefore, the capture of one or more terrorists rarely results in the compromised identity of the entire terrorist organi ation.

## Crime Prevention Measures

Although this publication is oriented to protecting Marines and their family members from terrorists, measures discussed in the following chapters are generally applicable to protection from crime as well. As an American, you must be careful when travelling at home or abroad. Crime is a more common threat to most military personnel than terrorism, however, the threat of terrorism has a far greater impact on national security. ou can narrow the chances of becoming a victim by increased awareness of potential problems and careful planning. racticing sound individual protective measures makes you a hard target.

# Chapter 2
# Protection Through Awareness

## Types of Targets

Terrorists prefer a target that involves little risk and a high probability of success. Terrorists evaluate a target's security profile, predictability, and value. The target's value is determined by its importance and possible benefits gained. nce a target has been evaluated by terrorists, the target is labeled in the terrorist's mind as either a soft or a hard target.

### Soft Targets

Soft targets are accessible, predictable, and unaware. They make it easy for strangers to access their private information e.g., phone numbers, addresses, schedules . Soft targets follow consistent routines at home and at work, allowing terrorists to predict a target's movements in advance. Soft targets are unaware of their surroundings and do not employ individual protective measures.

### Hard Targets

 ard targets are inaccessible, unpredictable, and aware. They make it difficult for terrorists to gain access to themselves or their families.  ard targets consciously vary their routines and avoid setting patterns in their daily life. They are security conscious, aware of their surroundings, and proactively adhere to individual protective measures.  ard targets do  **not**

- ut their names on mailbo es or e terior walls of their homes.
- Run or walk daily at the same time of day or to the same place.

- ash cars, mow lawns or have family cookouts the same day every week.
- Shop the same day of each week at the same store.
- Travel to and from home on the same route and at the same time of day.
- Attend church services at the same time of day and place each week.
- Sit in the same seat in a vehicle, restaurant, church, etc.
- Arrive at work, go to lunch, depart work at the same time of day every day.
- ick up the newspaper or mail at the same time of day every day.
- alk or feed the dog along the same route or at the same time of day every day.
- atroni e the same restaurants or bars or patroni e only American restaurants or bars.
- ark vehicles in the same area at church, social events, etc.
- arn the reputation of always lending a helping hand, e.g., aiding victims at staged roadside accidents.

## Identification of the Threat

earn about your destination the culture, language, local customs, and history of terrorist/criminal activity as soon as you know that you're going to be travelling outside the United States. Information is available from the following sources:

- Command's antiterrorism/force protection AT/F  officer or S-2/intelligence officer.

- Naval Criminal Investigative Service NCIS briefs and reports.
- U.S. efense Representative Force rotection fficer usually the U.S. mbassy's efense Attache or U.S. mbassy Regional Security fficer.
- Country handbooks from the Marine Corps Intelligence Activity MCIA .
- U.S. State epartment consular information sheets, public service announcements or travel warnings via the internet.
- ther Services' or other Government agencies' manuals and web sites. See Marine Corps rder MC 02.1C, *The Marine Corps Antiterrorism/Force Protection [AT/FP] Program,* for a listing of terrorism, law enforcement, and security information websites on the internet.
- Newspapers, maga ines, books, travel agents or tourist offices.
- eople who currently live or have lived in the area.

All commands are re uired to provide a evel I AT/F briefing to all Marines, civilian employees, and family members deploying or travelling outside the U.S. on official orders. evel I briefings include an area of responsibility A R specific threat briefing/ update for the area of travel. This brief stresses the need for a heightened awareness of the terrorist threat and reviews individual protective measures that can reduce individual vulnerability. See MC 02.1C for more details. nce in-country, additional information is available from the U.S. mbassy and the host country specifically:

- Are the terrorist groups in the area active
- Are the terrorist groups organi ing or reorgani ing

- hat are the local populace's attitudes towards the terrorist groups
- hat are the local populace's attitudes towards Americans
- oes the respective foreign government support, condone or condemn the terrorist activity
- hat is the potential for violence
- hat are the terrorists' methods of operation

Generally, when a terrorist group is successful with a certain method of operation, the group reuses it or it will be used by other terrorist groups. owever, ust because a terrorist group has not used a specific tactic in the past does not mean they won't develop new tactics or adopt similar tactics used by other terrorist groups.

## Visibility

Be alert to your surroundings, know and respect local customs and laws. on't call undue attention to yourself. Be unpredictable by varying the days and times of your activities and by varying routes you usually travel.

REMEMBER THREE BASIC RULES:

- BE ALERT
- KEEP A LOW PROFILE
- BE UNPREDICTABLE

Anyone who is highly visible is a potential, high-risk victim. ictims can be targeted for being an American, a very important person I , someone associated with I s or a target of opportunity.

## Identified as an American

ou can protect yourself from becoming a target if you avoid saying, doing, wearing, using, displaying or driving anything that readily identifies you as an American. ven if the local populace does not see Americans on a daily basis, global commerce and communications provides them access to maga ines, movies, television shows, and web sites that portray American lifestyles. The following paragraphs identify common indicators that easily identify Americans overseas.

### Uniforms

ear civilian clothes when traveling back and forth to work change into your uniform after you arrive at work and change into civilian clothes before you leave work.

### License Plates

Americans serving overseas may be issued different colored license plates or a different number or letter indicator on their license plates. If possible, use local license plates on any automobile driven. Avoid using vanity license plates or license plates with U.S. Marine Corps logos.

### Dress

Blend in with what the local populace or local tourist element wears. Flashy or trendy clothing can attract unwanted attention.

Clothes should not clearly identify you as an American for e ample, cowboy boots, American logo T-shirts, clothes bearing American sports teams, and e pensive athletic shoes.

## Speech

Although American dialect is hard to avoid, even if you speak the native language, avoid using military terminology and American slang.

## Customs and Habits

ven if you physically blend in with the local populace, your customs and habits can identify you as an American. If possible, you should adopt local or tourist customs and habits.

## Personal Behavior

Some Americans have the tendency to be loud and obno ious in the presence of the local populace. Another common mistake that Americans can make is to unnecessarily boast about American culture, wealth, technology, and military power, etc. in the presence of foreign nationals. Strive to blend in as much as possible, and not draw attention to yourself. **KEEP A LOW PROFILE**, especially in a public environment or with the local media.

## Tattoos and Jewelry

ear a shirt that covers tattoos with military or civilian slogans or logos when you go out. eave military ewelry such as service rings, medallions, and watches at home.

*Controversial Materials*

Avoid carrying potentially controversial materials such as gun maga ines, military publications, religious books, pornography or maga ines that can offend the local populace.

*Nationality Indicators*

American flags, decals, patches or logos easily identify you as an American. Avoid displaying them on your vehicles, clothes, in front of your home or place of employment.

*U.S. Government Bus Stops*

 o not wait for long periods of time at U.S. Government bus stops.  hen the bus approaches, walk toward the bus, stop short of the bus stop, and board the bus after the other passengers have boarded. Be especially observant for suspicious looking person-nel or ob ects such as unattended luggage or bo es.

*Currency*

 change a few U.S. dollars into the local currency before arriv-ing overseas. Use local currency and avoid carrying large amounts of money.

## Identified as Someone of Importance

Many people, including terrorists, e uate certain lifestyles with prominence. They believe that a prominent lifestyle is indicative of a person's importance to his government or company. Ameri-cans, in particular, are often treated by host governments as I s out of respect.  henever possible, avoid being treated as a I .

Avoid using your rank, title or position when introducing yourself or signing your name. Strive to maintain a low profile and blend in with the local populace. The issues identified in the following subparagraphs give the impression of importance and therefore should be avoided.

*Expensive Cars*

eople may think anyone who drives an e pensive car is important. Avoid driving e pensive vehicles. rive the type of vehicle that is common to the area in which you are located.

*Staff Cars*

eople may think that anyone driving around in a staff car from the American embassy must be important. Therefore, limit use of nonarmored staff cars.

*Bodyguards*

If you do not need bodyguards, do not use them. If you must have bodyguards, keep them to a minimum and ensure that they blend in with the other personnel around you they should not be obvious. nsure bodyguards pass a background check and are well trained.

*Chauffeurs*

Many people may believe that anyone who has a driver is a I . Therefore, perform your own driving if possible. If you do have a driver, the rear right seat is typically reserved for a I . Therefore, sit up front with the driver and occasionally rotate your seat position within the vehicle. ou should also

- nsure your driver has the re uired training so that he will not panic or free e in a high pressure situation.

- evelop an all-clear or distress signal e.g., a hat or cigarette pack on the dash between you and your chauffeur. A signal allows the driver to warn you of a problem prior to your approaching the vehicle.
- ave the driver open the door for you.
- Avoid giving your itinerary to your driver. All a driver needs to know is when and where to be. For e ample, you have the driver show up at 0 00, but you do not leave until 0 00. If possible, tell your driver your destination only after the car has started.

## Briefcases

In some countries, people think anyone carrying a briefcase is considered important. If possible, avoid carrying a briefcase unless it is the norm for the area. If the local populace uses backpacks, then you should also use a backpack.

## License Plates and Decals

If using diplomatic license plates, license plates with low numbers, or decals is unavoidable, employ proactive individual protective measures to reduce both vulnerability and visibility.

## Passports and Official Papers

iplomatic black and official red passports indicate someone of importance. Use a tourist blue or green passport whenever possible. If you use a tourist passport, consider placing your official passport, military I , travel orders, and related documents in your checked luggage. If you must carry official documents on your person, select a hiding place onboard your aircraft, bus, boat or train to hide them in case of a high acking. Try to memori e your passport number and other essential information to avoid flashing this information in front of other passengers. hile passing through customs, keep your passport out of sight by placing it in your airline ticket pouch. o not carry classified or official papers unless it is mission essential.

*Parking*

I s warrant their own parking spots usually very close to their offices, thus drawing attention to themselves and their importance. Therefore, avoid using a designated parking space  instead, park in an unmarked parking space and rotate where you park your vehicle.

*Domestic Employees*

In many foreign countries domestic employees such as maids, cooks, private guards, gardeners, and drivers are very affordable. owever, domestic help can provide terrorists with critical access to you and your family. If you are considering employing domestic help, ask for letters of reference and obtain a background check through the  mbassy, if possible.

- Avoid live-in domestic help. If they must have access to keys, never let them remove keys from the house.
- omestic employees should not allow anyone  including persons in police uniforms  to enter the house without permission from the family.
- Avoid providing transportation to and from work for any domestic employees. ay for a ta i or bus fare.
- If a domestic employee calls in sick, do not accept the temporary services of a relative  cousin  or sister .

- ave domestic employees report potential terrorist surveillance of your residence and watch for anyone loitering in the area or repeatedly driving or walking by.
- ay domestic help well and give cash rewards for following your security rules.
- Take special care to never discuss sensitive topics or detailed travel plans in their presence. Terrorists have successfully drawn this information from domestic employees in the past.

## Identified as a Target of Opportunity

These are the headlines that millions of Americans were viewing in une 19 when four Marine U.S. mbassy guards became targets of opportunity for the Faribundo Marti ara la ibercion Nacional FM N terrorist organi ation. These Marines were sitting outside of a very popular cafe in San Salvador when they were gunned down for being symbols of the

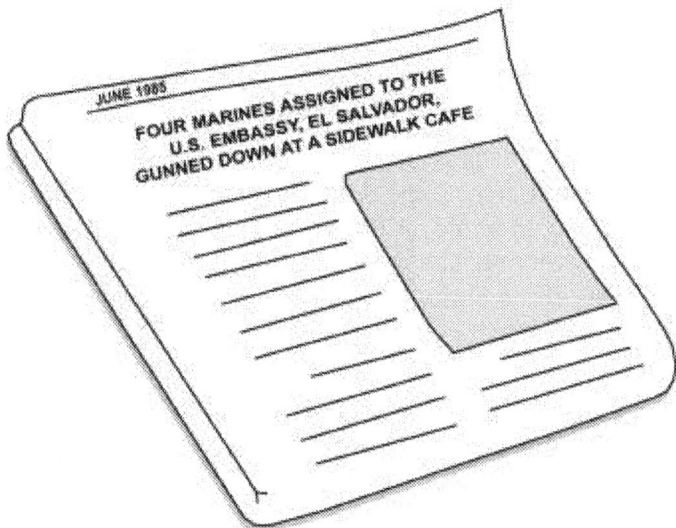

JUNE 1985
FOUR MARINES ASSIGNED TO THE U.S. EMBASSY, EL SALVADOR, GUNNED DOWN AT A SIDEWALK CAFE

United States. hen overseas, remember that you are a visual symbol of an American presence, values, prestige, and power. The longer you remain overseas, the more comfortable you may become. The more comfortable you become, the less you may think of yourself as a potential target. hile overseas, never allow yourself to become complacent. Safeguard information concerning yourself, your home, ob, and family. The more intelligence a terrorist can collect on you, the greater his chance of success. Terrorists gather their information from a variety of sources, which can include the following:

- arious internet sources, including command web pages.

- Aircraft loading manifests identify sending units, receiving units, departing facilities, and landing facilities.

- Bills of lading provide names of people moving into and out of an area.

- Immigration records provide names of people, dates of birth, and nationalities.

- Unit rosters provide names, addresses and phone numbers of individuals, spouses, and dependents. Unit rosters should be controlled and not posted in plain view.

- Manning boards provide an individual's name, duty position, s uad, platoon, etc. some even have photographs. If possible, offices should avoid using manning boards. If these are a necessity, they should be covered when not in use and kept in a locked office during nonduty hours. o not post them in front of the unit.

- Billeting offices often maintain a listing of all housing assign- ments. Security managers must ensure that billeting offices establish procedures to prevent unauthori ed disclosure of personal information.

- Telephone directories provide an individual's name, address, and phone number. If you must list your phone number in the telephone directory, re uest that only your name and number be included, not your address, rank or duty position.

- Some units or schools publish a  ho's  ho  book. If possible, avoid having your name listed in this type of publication.

-  uty rosters for the staff duty, drivers, military details, etc., should not be posted in plain view.  hen they become obsolete, they should be destroyed  not  ust thrown away .

-  iscarded mail or official correspondence can be used to identify an individual, the sender, and the place from which the correspondence was sent.  estroy any mail or official correspondence no longer needed and remove address labels from maga ines.

- The carbon from a credit card provides an individual's name and account number. Use the currency of the country you are visiting or working in. If you must use a credit card, also re uest the carbon copy.

- Checks can provide an individual's name, address, phone number, and social security number.  ave only minimal information printed on the front of your checks.

- Nameplates make it easy to find an individual in an office environment  avoid their use, if possible.

- Receipts from hotels, laundries, etc., identify an individual by name and often by room number. Consider using a nickname or an assumed name.

-  uggage should be generic and civilian in nature. Avoid displaying your rank, unit patches, decals, or any American identifiers on your luggage.

- Remove all destination and baggage claim tags from luggage as well as stickers, decals, and other markings that reveal that the luggage has been through U.S. Customs e.g., custom's stickers .

- Be aware of all the documentation that contains information about your unit, yourself, and your family. estroy all documentation, especially trash, that could be used by terrorists as a source of information.

## Family Members

Family members must be aware of the potential terrorist and criminal threat at home and abroad. They should also receive a evel I AT/F brief that emphasi es the importance of individual protective measures and physical security. nsure that your family members know and perform the following safeguards when traveling/living at home and abroad:

- The threat risk for the area.

- here they are at all times. A simple orientation to the area could prevent them from straying into dangerous areas.

- eep the house locked and secured whenever leaving the house. ercise caution upon return. Set up simple signals to alert family members or associates if there is danger.

- evelop and practice emergency procedures for use in the home such as:

  - vacuation due to fire.
  - Intruders in the residence upon arriving home.
  - Intruders breaking into the house.

- ocation and phone numbers of the U.S. mbassy, military base, neighbors, and all emergency services such as police, fire department, and medical services, and other safe locations for refuge or assistance.

- Carry small cards with emergency phrases in the respective foreign language, and post these phrases by the telephone.

- In preparation for emergencies, maintain survival items e.g., supply of fresh water, nonperishable food, candles, lanterns, flashlights, e tra batteries, blankets, portable radio, camping stove with spare fuel, a e, first aid kit, and other appropriate items . Consider maintaining a similar kit for your car for emergency situations in isolated areas. For more information see MCR -02F/FM 21- , *Survival.*

- Take an ample supply of medications that family members use. Also keep a copy of the prescription, statement from a physician, and know the generic name of the medication so you can reorder it abroad. Also, keep eyeglass prescriptions on hand.

- Always carry identification documents. Carry a card stating blood type and allergies to particular medications. The card should be bilingual/multilingual nglish and the host nation language s .

## Special Precautions for Children

idnapping is a potential tool used to e tort ransom money that finances terrorist organi ations or may be used as an attempt to force you to assist in a terrorist operation. Special precautions include the following:

- Never leave children alone or unattended. eave children only with responsible and trustworthy individuals capable of handling emergency situations.

- Instruct children to keep doors and windows locked, and never to allow strangers into the house. iscourage children from answering the door, especially during hours of darkness.

- If possible, locate children's rooms in areas not easily accessible from the outside.

- Instruct children to never leave home without telling the parents. They should only travel in groups and avoid isolated areas especially when travelling to and from school. Accompany young children to and from bus stops, where necessary.

- Children should only use locally approved play areas where recreational activities are supervised by responsible adults and where police protection is readily available.

- Children should refuse automobile rides from strangers and refuse to accompany strangers anywhere on foot, even if a stranger says, our Mom/ ad sent me and said it was ok .

- Inform school authorities to never release children to any person who is not a family member. Instruct children to call home if a stranger is there to pick them up.

- Children should be told to refuse gifts from strangers and to avoid providing information to strangers such as their name and where they live.

- Children should immediately report anyone who attempts to approach them to the nearest person or authority  teacher, police .

- Instruct children not to discuss what you do and tell them to inform you if they are uestioned about you by anyone.

## Home Physical Security

Criminals remain the most likely threat in your home. owever, terrorists have conducted operations at the homes of servicemen overseas. The following section provides some basic information to make your home a hard target. evelop a security plan that includes the following:

- **Operations Security.** on't provide information to potential terrorists or criminals via the mail, phone, computer or trashcan.
- **Outer Security.** Use available assets  local shop owners, neighbors, domestic employees, guards, family etc. to detect potential surveillance.
- **Inner Security.** stablish a warning system with pets, alarms, and motion sensors.
- **Barriers.** Fences, walls, locked doors and windows, secure rooms to go to in an emergency.
- **Communications.** hone, cell phones, megaphones, intercoms, radios, audible alarms, linked security systems.
- **Deterrent/Response Systems.** Guards, pets, weapons  if authori ed , and fire e tinguishers.

### General

- Change or re-key locks when you move in or when a key is lost by a family member. Maintain strict control of all keys. Change the security code in the garage door opener. Never leave house or trunk key with your ignition key while your car is being serviced or parked by an attendant.

- on't open doors to strangers. bserve them through a peephole viewer. stablish procedures for accepting deliveries such as: verifying identities of delivery person, checking the identity of the deliverer with the appropriate dispatcher, refusing all une pected packages.

- Allow maintenance work only on a scheduled basis. Unless a clear emergency e ists. Be alert to people disguised as public utility crews, road workers, vendors etc., who might station themselves near the house to observe activities and gather information.

- Note parked or abandoned vehicles near the entrance or walls of the residence.

- Make residence appear occupied while you are away by using timers to control lights, T s, and radios.
  - Ask neighbors to ad ust blinds and draperies and pick up newspapers and mail.
  - Schedule regular lawn work.
  - Notify local law enforcement or military police if you will be away for an e tended period.

## Residential Physical Security

- Routinely keep all doors, skylights, roof doors, and windows locked. eep all window curtains and blinds tightly closed after sundown.

- Install lighting all around the house and yard link to timers and sensors.

- nsure door frames, doors, locks, and bolts are of solid construction. nsure door hinges e posed to outside of house are pinned or spot-welded to prevent removal of the hinge bolt.

- nsure fuse bo es are secure from tampering.

- Remove all trees, poles, ladders, etc., that might help an intruder scale fences, walls or gain access to second floor windows. Remove dense foliage or shrubbery near gates, garages, windows or doors that might conceal an intruder.

- Install intrusion detection, smoke, and fire alarms. nsure intrusion detection alarms covers both the perimeter doors and windows and interior motion and/or glass break sensors . ave the alarms monitored through a reputable security service or police. Train family members to use and test alarms regularly.

- If possible, select and prepare an interior safe room for use in case of emergencies. The safe room should have a sturdy door with a lock and an emergency e it, if possible. Bathrooms on upper floors are generally good, safe rooms.

- Store emergency and first aid supplies in the saferoom. Bars or grillwork on saferoom windows should be locked from the inside to e pedite escape.

- eep keys to locks, a rope or chain ladder to ease escape, and a means of communication e.g., cellular phone and radio transmitter.

## Telephones

- on't place your name in a public local phone directory.

- If you receive obscene, threatening or annoying phone calls or an unusual number of wrong or silent callers, report this to the police. Use caller-I or call block, if available.

- Answer the phone without providing any personal information. Be especially cautious when sending personal information over computer on-line services.

- Report any interruption or unusual interference with phone, electrical or computer service. This could be the first indication of bugging your phone line.

- eep a cellular phone charged and available, particularly at night.

## Letter Bombs and Biological Mailings

eightened personal security involves treating any suspicious-looking mail letter or package as a bomb or a potential biological threat. If you think any mail is suspicious, contact the military police or appropriate security officials and let them investigate. o not attempt to handle the mail yourself. ou should e amine your mail for the following suspicious features:

- It is from a stranger or an unknown place
- Is the return address missing
- Is there an e cessive amount of postage
- Is the si e e cessive or unusual
- oes it have e ternal wires or strings that protrude
- Is the spelling correct
- oes the return address and place of postmark match
- oes the handwriting appear to be foreign
- oes it smell peculiar
- Is it unusually heavy or light
- Is it unbalanced lopsided

- Are there any oily, sticky or powdery substances on the outside of the letter or package
- oes it have springiness on the top, bottom or sides

ou should use the following guidelines if you suspect that a piece of mail contains a bomb or biological agent:

- on't panic.
- o not shake the empty contents of any suspicious envelope or package. If any powder or substance leaks out, do not attempt to clean it up.
- lace the envelope or package in a plastic bag or some other type of container to prevent leakage of the contents. If you do not have a container, cover the mail and do not remove the cover. If powder or any other substance has already leaked out, cover that also. ou can cover the mail with clothing, paper, trash cans, etc.
- eave the room and close the door. Secure the area to prevent others from entering.
- ash your hands with soap and water to prevent spreading any biological agent to your skin or respiratory system.
- Report the incident to authorities. If at home, dial 911 and report the incident to your local law enforcement agency. If at work, report the incident to the governing law enforcement agency and notify your building security official or an available supervisor.
- ist all of the people who were in the room or area when the suspicious mail was recogni ed. Give this list to both the local health authorities and law enforcement officials.

All mail that is sent overseas should be delivered via Army/Fleet ost ffices or through the U.S. mbassy to allow for proper bomb detection and inspection by trained mail handlers.

## Vehicle Bomb Search

A large number of terrorist attacks take place in or around a vehicle, typically by some sort of e plosive device. This occurs because bombs are relatively easy to make and plant on e posed and unattended vehicles. ou need to learn how to search a vehicle for tampering and to recogni e danger signs. Appendi A contains procedures and tips for conducting a vehicle bomb search. By routinely inspecting your vehicle, you give the impression of being a hard target.

## Travel

Traveling is one of the most opportune times for a terrorist attack. ou are the most vulnerable and predictable in the morning as you enter or leave your uarters, your place of work, or your vehicle. our understanding and application of the following information can reduce your chances of becoming a victim of a terrorist attack while traveling:

| General Precautions |
|---|
| Remain alert; travel in groups or pairs in well-lighted, busy areas. |
| Watch your luggage at all times. Use concealed bag tags. |
| Establish alternate routes from each starting place to each destination. Make sure at least one person you work with and someone in your family are aware of these routes and the approximate time it takes you to travel these routes. |
| Keep travel arrangements confidential as much as possible. |
| Avoid using rank or military addresses on tickets, travel documents, and hotel reservations. |
| Make a copy of the following and place in different pieces of luggage: passport, ID card, and official papers. If lost or stolen, these items can be replaced at a U.S. Embassy, Consulate or military facility. |
| Register with the U.S. Embassy upon arrival in country either in person or via phone. Carry a card that has the location and phone number of U.S. military facilities and the U.S. Embassy and Consulates in the area. These are vital safehavens during emergencies. |
| Maintain a low profile. Do not discuss your U.S. government affiliation with any other passengers. |
| Get a detailed briefing from the Force Protection/S-2/intelligence officer on the cities you plan to visit. The briefing should include the threat, the safest routes to use, safehavens, areas to avoid, and anything else pertaining to your mission and safety. |
| Avoid using public transportation. Buses and trains are preferred to a taxi. If you must travel in a taxi, specify the route you want the taxi driver to take and look for the photo identification or license to ensure that the photo matches the driver. |

Learn common phrases and greetings and how to ask for assistance or help in the local language.

Know how to use public phones and carry enough change (in the local currency) to make a phone call. Calling cards can be used in many countries also.

Learn the names and phone numbers of persons to contact at your destination, including emergency numbers.

### Safeguards while Driving

Park your car for easy escape (pointed outwards).

Lock your car and garage when you park overnight. Alternate use of parking garages if possible. Park in well lighted areas if you must park on the street.

Walk to your car with keys in hand, ready to use.

Perform a quick internal and external check of car. See appendix A for vehicle bomb check.

Start your car immediately after conducting your vehicle bomb search. Do this before you adjust your seat or mirrors. You should be prepared for rapid escape if necessary.

Lock your doors and keep your windows up.

Wear your seatbelt.

Avoid traveling alone and during late hours. Know where the dangerous areas in the city are and avoid them.

Travel only on busy, well-traveled thoroughfares, especially routes that allow speeds over 25 mph. Most attacks occur in stop and go traffic. Avoid one-way streets and other choke points such as bridges, traffic circles, and narrow alleyways. Avoid isolated secondary roads.

| |
|---|
| Enter and exit your vehicle at busy locations. |
| Know en route safehavens such as police and fire stations, military posts, and checkpoints you can drive to. If you feel you are being followed, don't go directly home. |
| Avoid carrying classified material. If driving, lock classified materials in the trunk. |
| Varying times and routes driving to and from work. |
| If possible, use different building entrances and exits. |
| Keep your vehicle in good mechanical condition and your gas tank at least half full. Ensure you have a locking gas cap. |
| Keep safety equipment (e.g., cellular phone, fire extinguisher) inside your vehicle in good working order. Consider carrying a survival kit. (See MCRP 3-02F/FM 21-76, *Survival*.) |
| Avoid driving close behind other vehicles or in any situation where you can get boxed in or forced to a curb. Have an evasive plan ready. Sometimes making a simple U-turn is enough to get you out of danger. |
| Keep at least one-half car length of empty space in front of your vehicle when stopped at traffic signals and stop signs. This gives you room to escape in a kidnapping or armed attack/assassination attempt. |
| Never pick up hitchhikers. |
| In an emergency, drive on flat tires until reaching a well-lighted, well-traveled area or safehaven. |

In the event of mechanical failure, set out warning triangles/ flares, raise the hood, activate emergency flashers, and stay inside. If someone stops to offer assistance, ask them to notify the police or road service. If you feel unsure of the situation, don't get out of the car until the police or road service arrives. If you feel threatened by strangers, stay in the car with the doors locked. Use vehicle's horn to attract attention.

### Safeguards while Walking

Be alert to the possibility of surveillance. Before leaving a building or mode of transportation, check up and down the street for suspicious looking cars or individuals.

Walk facing traffic at all times.

Walk on the center of the sidewalk, this allows you to see around corners. Walking next to the street affords someone the opportunity to push you out into the street.

Remain alert when walking across alley entrances or other places where a terrorist could be hiding.

Walk only in lighted areas. Avoid bad sections of town.

Avoid walking in noisy areas; e.g., a construction site.

Stay near people. Don't walk in isolated areas; e.g., alleys.

Avoid hostile crowds by turning back or crossing the street.

If you suspect you are being followed, move as quickly as possible to a safe-haven (e.g., police station or government office).

### Safeguards while Flying

If possible, buy your ticket at the last possible moment to prevent unauthorized personnel from finding out about your travel plans.

| |
|---|
| Choose flights that will route you through an airport with a history of good security measures. |
| Avoid countries, airports or airlines that are currently targets of terrorist organizations. Direct flights are best. If possible use military air, military charter or U.S. flag carriers. |
| Arrive early. Don't loiter near the ticket counter, luggage check-in or security area. Go through security as quickly as possible to the boarding area. Only use shops, restaurants, and lounges in the security area, not the main terminal. |
| Buy your ticket at a travel agency that offers you seat selection and gives you a boarding pass when you buy your ticket. Ask for a window seat near the center of the aircraft. Terrorists generally select passengers for abuse that are sitting in more easily accessible aisle seats. |
| Don't let your carry-on luggage out of your sight and don't agree to "watch" someone else's luggage. |
| Keep your eyes open for any suspicious activity such as an individual who gets up and leaves behind bags, packages, etc. If you see something suspicious, get out of the area quickly and report it to airport security officials! |
| Stay within the restricted or boarding areas of the airport, or leave the airport if possible or practical when you have a long layover for several hours. |
| No matter where you are in the terminal, identify objects suitable for cover in the event of an attack. Pillars, trash cans, luggage, large planters, counters and furniture can provide some protection. |
| Sit with your back against a wall, facing the crowd to give you greater awareness to your surroundings. |
| Avoid seats in first class. |

Count the number of seats to the closest emergency exit so that you will be able to find your way out in case the lights go out, or if the compartment fills with smoke.

Avoid telling other passengers that you are in the military or otherwise confiding in them. On a foreign carrier, avoid speaking English as much as possible.

Inform someone of your destination and get in the habit of checking in with them before you depart and after you reach your destination. This could provide authorities with a starting point if you should become missing.

At the first indication of a hijacking, hide all documents, identification cards, and official passports that could identify you as military.

### Safeguards while Staying in Hotels

Stay at DOD facilities whenever possible for security.

Request another room if one has been reserved for you. Do not give your room number to strangers.

Avoid street-level rooms. Ask for a room between the second and eighth floors. This puts you high enough to avoid easy access from the outside and still be low enough for local fire equipment to reach.

Check before exiting from an elevator or your room for objects that seem out of place or for strangers who seem to be loitering.

Answer the hotel phone with hello, not your name.

Never answer hotel paging. If you are expecting someone, go to the lobby, but don't go to the desk and identify yourself, check to see if the caller is whom you are waiting for.

Keep your room key on you at all times. Don't leave a copy of your room key on your key chain for the parking attendants.

| |
|---|
| Be careful answering the door. First, check to see who it is through the peep hole or side window and arrange knock signals with your traveling companions. |
| Watch for anyone loitering in halls, lobbies, or public areas or for anyone carrying objects that could be used as a weapon. |
| Vary your arrival and departure times. |
| Vary how you enter and exit the building; e.g., use a hotel's entrance as well as its elevators and stairwells. |
| Know where emergency exits and fire extinguishers are located. |
| Avoid frequent exposure on windows and balconies. Keep your room draperies closed. Conduct business in your room, not in the lobby or hallways. |
| Inspect your room thoroughly upon entering. Keep your room and personal effects neat and orderly. This practice helps you recognize tampering or strange, out-of-place objects. |
| Place a piece of tape on the door crack or a string in the door jam. If it has moved while you were out, you will know that someone has entered your room during your absence. |
| Lock the door and use the chain. |
| Place the DO NOT DISTURB sign on the door. |
| Avoid maid service and never admit a stranger to your room. |
| Consider purchasing a portable door alarm, this will awaken you if someone attempts to enter while you are sleeping. |
| Place a large screw into the space between the door and the door frame, this will delay anyone's entry into the room. |

| |
|---|
| Leave the lights, television or radio on when you are out of the room to give the appearance that someone is still there. |
| Find out if the hotel has security guards; if so, determine how many, their hours of duty, equipment they use, their expertise, and how to locate them by phone and in person. |
| Do not discuss travel plans over hotel phones. The lines could be "bugged." |
| Do not take the first taxi in line when leaving your hotel and don't allow strangers to direct you to a specific cab. |

## Detecting Surveillance

Terrorist operations are normally meticulously planned, allowing for the greatest chance of success and safe escape for the terrorists. Reducing vulnerability with security enhancements is vital to your efforts to deter terrorist attacks. ually important is surveillance detection. In most cases, the target that terrorists select to attack is based on lengthy surveillance. Through surveillance, they hope to learn about your habits and assess where you are vulnerable. By practicing good individual protective measures, you not only disrupt their intelligence gathering efforts, but you also make yourself a hard target. Terrorists want to hit soft targets, which minimi es their risk of failure. In cases of targets of opportunity, however, the surveillance may last only for a few minutes to hours to confirm the ease of the target. owever, terrorists will usually abandon hard targets and move on to another soft target.

Upon arrival in a new area, begin determining what is normal and routine. nce you've determined what is normal and routine, it is easier to determine what is unusual. This makes the problem of identifying surveillance simpler.

ften initial surveillance efforts are conducted by less e perienced personnel who may often make mistakes. For e ample, terrorists will often show up at a surveillance location immediately prior to their target's arrival and depart immediately after the target leaves. A surveillance program involving family members, neighbors, and domestic employees can often detect this surveillance.

ook for people who are in the wrong place or dressed inappropriately. liminate stereotypes about terrorist surveillance personnel they are often women and children. Be particularly observant when traveling to and from your home or office. ook up and down the streets for suspicious vehicles, motorcycles, mopeds, etc. Note people near your home or place of work who appear to be repair personnel, utility crews, or even peddlers. Ask yourself if they appear genuine or is something unusual

## Types of Surveillance

- Stationary - At home, along route or at work.
- Following - n foot or by vehicle.
- Monitoring - Telephone, mail, computers.
- Searching - uggage, personal effects, trash.
- avesdropping - lectronic and personnel.

Terrorists sometimes employ an elaborate system involving several people and vehicles. Typical surveillance vehicles are motorcycles and cars with multiple personnel. Become familiar with local vehicle makes and models. Memori e and write down license plate numbers. etermine if a surveillance pattern is developing.

## Surveillance Indicators

- Illegally parked or occupied parked vehicles.
- Cars with large mirrors.
- Cars that suddenly pull out of parking places or side streets when you pass, cars that move with you when you move, or cars that pass you and immediately park.
- Cars slowly maneuvering through turns and intersections or vehicles signalling for turns but do not turn.
- Flashing lights for signaling between cars.
- Unusual speeding up, slowing down or running red lights to stay up with you.

Conduct a route analysis of your principal routes that you make on routine trips. Identify chokepoints where your vehicle must slow down. Typically these chokepoints are: traffic circles, one-way streets, bridges, and ma or intersections. Search out safe-havens that you can pull into along the route in the event of emergency. If you think you're being followed, go directly to a safehaven, not your home. Safehavens are generally well lit, public facilities where persons will respond to your re uest for help.
   amples of a safehaven might be a police station, fire station, large shopping mall, busy restaurant.

If you are aware of surveillance, never let those watching you know you have figured out what they are doing. Never confront them. Terrorists and criminal elements are typically armed, don't want to be identified, and may react violently in a confrontation.

## Reaction (if in a Vehicle)

- Circle the block for confirmation of surveillance.
- o not stop or take other actions that could lead to confrontation
- If possible, get a description of the car and its occupants.
- Go to the nearest safehaven. Report incident to the nearest security or law enforcement organi ation.

## Reaction (if on Foot)

- Move rapidly towards a safehaven avoiding any route you routinely use.
- If a safehaven is not immediately available, move into a crowded area.
- Immediately report suspicions to nearest security element or local law enforcement.

# Attack Recognition

If terrorists succeed in surveilling you and plan an attack, the ne t place to foil their efforts is to recogni e their intentions and prepare to escape. Recogni ing an attack scenario is difficult. ften what may appear to be an attack is more likely to be innocent circumstances. owever, alertness and willingness to act are the keys to surviving a genuine attack scenario.

## Abnormal Situations

- Individuals who appear to be e cessively nervous and seem out of place by dress or mannerisms.
- Individuals wearing unusually long or heavy clothing for the environment.

- Individuals who appear to be acting as lookouts along your route of travel.

- ehicles that hit your car from the front or rear.

- Unusual detours, vehicle roadblocks, cones, or other barriers. Be prepared to escape by going around the obstacle or ramming it.

- ehicles traveling with items protruding from side doors or vans traveling with side doors open.

- isabled vehicles, hitchhikers or distressed accident victims seeking your assistance are commonly employed traps.

- A flagman, workman or fake police or government checkpoint stopping your car at a suspicious place.

- Sudden unusual activity or the une plained absence of local civilians.

- Gunfire.

## Escape, Evade or Confront

nce you recogni e an attack is occurring, decisions must be made immediately. If the scenario is an armed attack or assassination attempt, get out of the kill one. Typically terrorists have a relatively narrow window of time and may have restricted fields of fire due to obstacles in their path. nce you e it the kill one, terrorists will rarely pursue you since they must begin their own escape and evasion plan. In emergency situations, it may be a matter of survival to employ evasive driving techni ues in order to arrive at the nearest safehaven. Use of evasive driving techni ues may also be to you advantage by attracting the attention of local law enforcement. If on foot, take advantage of the density of crowds and layouts of buildings to evade pursuers. hen you feel you have evaded the terrorists and are out of immediate danger, contact security forces or law enforcement for assistance.

In some cases, you may become captive as were the passengers on board the ill-fated flights of 11 September 2001. scape and evasion were not possible. The only chance for those passengers to survive was to confront the terrorists in order to regain control of the aircraft. n one aircraft, although the plane crashed killing all on board, the passengers' confrontation with the terrorists saved countless lives because the aircraft never reached its intended target.

## Incident Reaction

### Bombs

Should a bomb e plode outside the building, do not rush to the window to see what happened. Immediately seek cover in a protected area due to the possibility of a secondary, probably larger e plosion  referred to as a double bombing . Terrorists may use an initial bomb to breach outer security, then a second bomb on the target, and may follow-up the bombing with an armed attack. In a variation, terrorists can place an initial bomb, followed by a second bomb shortly thereafter to kill or in ure security forces and emergency services responding to the initial bomb.

In the city, if you are on the street when a terrorist bomb e plosion occurs,  uickly get inside the nearest building and remain there. Shattered glass and other debris from high-rise buildings can fall for blocks around the point of e plosion. As soon as practical, following a bombing:

- Notify the proper authorities.

- vacuate the wounded based on the situation. o not impede the efforts of emergency services. itnesses to the bombing will naturally approach the e plosion area to aid in searching for casualties. Authorities will also be trying to coordinate the search and will want to limit the number of searchers due to the threat of additional e plosions and secondary effects such as falling masonry or fires.

- Move to a clear area, away from ob ects such as automobiles, buildings, and garbage containers.

## Armed Attack or Assassination

If in an office or hotel, uickly lock the door, turn out the lights, grab the telephone, and get down on the floor. Call building security immediately. Telephone connections outside the building or hotel might be difficult to obtain. If no security office is available, call the local authorities. Tell the authorities e actly what you heard and provide them with the address, building, floor, room number, and telephone number. Stay in a protected area, and if possible, take the phone with you. If you believe you are involved in a terrorist takeover, hide your wallet and identification.

## Arsons and Firebombings

 ercise normal fire safety precautions. owever, do not gather in open areas such as parking lots or areas where others are congregating. Terrorists could stage an arson attack or false fire

alarm to get a crowd out of a building and then conduct a bombing or armed attack.

## Hijackings, Skyjackings, and Kidnappings

Reactions to these and similar attacks are described in chapter .

## Additional Individual Aids

Appendi B contains the complete April 2000 te t of oint Staff
Guide 2 0, *Service Member's Personal Protection Guide: A
Self-Help Handbook to Combatting Terrorism While Overseas.*
Appendi C and  are wallet-si ed cards containing pertinent
individual protective measures.

For detailed checklists and discussions of Antiterrorism Individual rotective Measures, see    irective 0-2000.12- ,
*Protection of DOD Personnel and Activities Against Acts of Ter-
rorism and Political Turbulence.*

# Chapter 3
# Hostage Survival

## The Hostage

ostage-taking is a way for terrorists to achieve a bargaining position by forcing a confrontation with authorities. It will remain an effective terrorist tool as long as mankind values human life. ope-fully, you will never become a hostage, but if you do, knowing how to react will improve your chances of survival. our role as a hostage is to *survive with honor* not to kill the terrorists or get you or your fellow hostages killed. Remember, most hostages survive a hostage-taking. Terrorists select hostages for a variety of reasons. The hostage may have a prominent ob or social status or may be

- ell known, so that terrorists receive widespread media attention.

- An American.

- ated by the terrorists, or the terrorists may blame the hostage for any setbacks they have suffered from their own government's forces. For e ample, U.S. military advisors in l Salvador were despised by terrorists of the FM N because of the assistance the advisors provided the l Salvadorian government.

- aluable to employers and families for e ample, families and civilian firms have paid ransoms to secure a hostage's release.

- Seen as a threat to the terrorists. For e ample, in Colombia, the terrorist groups M-19 and Revolutionary Armed Forces of Colombia FARC , who make more than 100 million each year from cocaine sales, target special agents of the U.S. rug nforcement Administration.

These reasons identify why a particular person may be targeted for a hostage-taking. But, in most cases, a hostage is an innocent victim of circumstances someone who was in the wrong place at the wrong time. Therefore, you must prepare yourself to respond both mentally and physically to a hostage-taking incident.

---

### Terrorist Suicide Missions

If under the control of terrorists, you must rapidly determine the terrorists' intent: to establish a bargaining position and elicit publicity or to carry out a suicide mission. The passengers on board the planes used in the 11 September 2001 attacks were not hostages in the traditional sense because the terrorists never intended to use them to achieve a bargaining position. Those passengers were simply on board a skyjacked aircraft that was being used for a suicide mission. In such terrorist suicide mission situations, the techniques described in the "Escape or Surrender" section on page 3-8 may be the most useful.

---

## The Hostage-Takers

The following paragraphs address broad categories of hostage-takers the ones that are the norm. The lines between the categories may blur or overlap, and the hostage-taker may move from one category to another based on a goal. Multiple subsets may also e ist within each category.

## Political Extremist

Most hostage-takers are political e tremists. They typically operate within a military-type structure. Their operations are usually
well planned. They typically resist appeals based on morals,
decency or fear for their own safety. They are often prepared to
die for their cause.

Statistically, leaders of political e tremist groups are single,
urban, bright, and dedicated to their cause. They are often col-
lege graduates with professional backgrounds. They often come
from upper or upper middle class families whose parents are
politically active but not violent. They tend to be abnormally
idealistic and infle ible.

## Fleeing Criminal

Fleeing criminals take hostages on impulse, typically to avoid
immediate apprehension and to have a bargaining chip for escape.
Authorities must handle a fleeing criminal with caution. If he
feels a sudden loss of power, it can create agitation, despair, or
panic. ith these emotions at the forefront, he also may impul-
sively kill a hostage. Therefore, time and patience in dealing with
the hostage-taker is critical. The fleeing criminal will often settle
for much less than originally demanded if he perceives that he is
slowly losing power, control of the situation, or facing death.
Many times, he will surrender if allowed to give up with dignity.

## Wronged Person

A hostage-taker who feels he is a wronged person is motivated by
personal revenge. e seeks to notify society of the defects in the
system or the establishment. e attempts to effect ustice in order

to right a wrong or to publici e what he feels is an in ustice. The hostage may represent the  system  to the hostage-taker  if so, the hostage could be in increased danger. This type of hostage-taker is convinced that he is absolutely right in his behavior.  ften, gentle persuasion is re uired to convince him that he needs to end the situation and release the hostage.

## Religious Extremist

 ealing with hostage-takers who are religious e tremists re uires time, patience, and sensitivity. Religious e tremists share a common, unshakable belief in the righteousness of their cause. They may perceive that their source of power comes from their god or the leaders of their cult or group. They may see themselves as superior to others simply because of their beliefs. Individuals who  oin cults or radical religious groups often lack personal confidence and  oin these types of organi ations to bolster their self-esteem.

Religious e tremists may feel that they must succeed or die for their faith. Some religious cults and groups believe that to die at the hand of the nonbeliever is the holiest achievement possible. This way of thinking greatly increases the threat to the hostage. The hostage may also be seen as a  sacrificial lamb,  one who must die for the sins of others.

## Mentally Disturbed

A hostage-taker who is mentally disturbed is not normally associated with an organi ed terrorist group.  owever, this type of hostage-taker conducts over half of the hostage-taking incidents. Usually, the mentally disturbed hostage-taker acts alone. Authorities may have difficulty establishing and maintaining a rapport with the mentally disturbed hostage-taker. If challenged or threatened by authorities, the mentally disturbed hostage-taker may easily accept the murder of a hostage, his own personal suicide or both.

## Personal Contingency Planning

As a Marine and a representative of the U.S. Armed Forces, you are a potential hostage even in low and negligible threat areas. ou must prepare for your own personal safety, prepare your family, and prepare for the potential of becoming a hostage and the ensuing captivity.

Although nothing can fully prepare you for the e perience of becoming a hostage, knowing that the following issues have been addressed will lessen the trauma on you and your family:

- our will is current and your family knows its location.
- Appropriate powers of attorney are given to a spouse, relation, or trusted friend.
- our family knows who to contact for assistance.
- Family finances are settled so they do not suffer financially during your captivity. Family members should have access to money, airline tickets, credit cards, insurance policies, etc.

All family members should assemble a personal history/information sheet, preferably in their own handwriting, which can be used as an aid to law enforcement and intelligence officials in the event of an incident. This sheet should contain the following information items:

- Name and nicknames.
- lace and date of birth.

- ome address and telephone number.
- Secondary address and telephone number
- recise physical description e.g., height, weight, scars, tattoos, dentures, etc. .
- ther identifying characteristics e.g., birthmarks, physical handicaps .
- rescription for eyeglasses used.
- Special medications and instructions for their use.
- ehicles types and license .
- School type, class, address, teachers .
- Recent information on educational ualifications and hobbies.
- Information about friends residing in diverse locations and their phone numbers.
- reparing a brief family member oral history on a cassette recording may be helpful. This can be used to help identify voices on recordings mailed to authorities in the event of a kidnapping. Samples of handwriting taken under various conditions e.g., writing on the hood of a car, writing on bare ground , may also be helpful.

our family also needs to know how to react if you are taken hostage. Typically, terrorists carry radios so they can listen to the news and monitor the world's reaction to the hostage-taking and to receive further instructions from their superiors. our family should not grant interviews to the media. If confronted by the media for a statement, your family should say that they hope the terrorists will release the hostages, and that the ordeal will be over soon. ou should advise your family not to e press fear for your safety, identify you as a Marine, or provide the terrorists with any information that will help the terrorists select you as a victim or endanger the lives of the other hostages.

## Department of Defense Directive 1300.7

Closely related to personal contingency planning is developing a good understanding of the policy and guidance contained in irective ir 1 00. , *Training and Education Measures Necessary to Support the Code of Conduct.* The si articles of the Code of Conduct outline basic responsibilities and moral obligations of members of the U.S Armed Forces. The code is designed to assist military personnel, both in combat and those being held as prisoners of war, conduct themselves in a manner that brings credit upon themselves and their country.

hen military personnel participate in military operations other than war or are assigned overseas, they are potentially sub ect to detention by hostile governments or captivity by terrorist groups.
 irective 1 00. builds upon the spirit and intent of the Code of Conduct to provide guidance to servicemembers that specifically applies to peacetime detention or captivity. ey elements of the directive include the following:

- Maintaining faith with fellow hostages by communication and re ecting privileges and special favors.
- Resisting e ploitation for information or propaganda purposes.
- Maintaining proper bearing and displaying courtesy.
- Resisting disclosure of classified information or materials.
- rgani ing in a military manner to the fullest e tent possible.
- Avoiding embarrassment to the U.S. and host governments.

- Authori ing acceptance of release unless doing so re uires you to compromise your honor or causes damage to the U.S. government or allies.

- Authori ing escape attempts, if you view such as attempt as your only hope.

- roviding specific guidance and legal considerations involving detention by hostile governments.

## Escape or Surrender

uring the initial moment of capture, you must make an instant decision escape or surrender. ven though it is the most dangerous time of a hostage ordeal, you must remain calm. o not make any sudden movement that may rattle an already an ious gunman. Abductors are tense adrenaline is flowing. Terrorists themselves feel vulnerable until they are convinced they have established firm control over their hostages. Unintentional violence can be committed with the slightest provocation. For e ample, do not make eye contact with the captors initially. Be polite and cooperate. ou may need to reassure your abductors that you are not trying to escape by controlling your emotions, following instructions, and avoiding physical resistance.

Terrorists meticulously plan to capture hostages. Initiative, time, location, and circumstances of the capture usually favor the captors, not hostages. owever, the best opportunity to escape is normally during the confusion of the takeover while you are still in a relatively public place. uring this period the hostage takers are focused on establishing control and may leave openings for escape. If you decide in advance to try to escape, try to plan and practice doing so. Mental alertness improves the chances of escape. hile waiting for an opportunity to escape, continue passive information collection on

- Appearance, accents, rank structure, e uipment, and routines of the terrorists.
- Strengths and weaknesses of the facility and its personnel.
- Conditions and surrounding area that could impact an escape attempt.
- Items within the detention area that can be used to support an escape effort.

 scape from detention by terrorists is risky but may become necessary if conditions deteriorate to the point that the risks associated with escape are less than the risks of remaining captive. These risks would include the credible threat of torture and death at the hands of the terrorists.  scape attempts should be made only after careful consideration of the risk of violence, chance of success, and possible detrimental effects on hostages remaining behind.

If you eliminate escape as an option, avoid physical resistance. Assure your captors of your intention to cooperate fully.

## Intimidation and Control

Remember, hostage-takers usually want you alive  They may use drugs, blindfolds, or gags when they abduct you, but try not to be alarmed or resist unduly. If you struggle, hostage-takers may resort to more severe measures of restraint.  ostage-takers use blindfolds or hoods to keep you from knowing where you are being taken, as well as to prevent you from identifying them later.  ou should not attempt to remove a blindfold or hood, if you see your abductors they may kill you. ikewise, you should not attempt to remove an abductor's mask or hood if they are wearing one.

ostage-takers may also drug their victims, usually at the beginning of an operation, to make the victim sleep and keep him pacified. This e perience should not be alarming. At this stage, your life is almost as important to the hostage-taker as it is to you. rugs used to put you to sleep do not have lasting side effects. If hostage-takers should use drugs such as heroin, lysergic acid diethylamide S , or sleeping pills, you can typically recover from this uicker than you can from physical abuse. ostage-takers may use truth-serum drugs, but these drugs are typically inefficient, and their results are similar to the consumption of too much alcohol.

## Stabilization

If you are abducted, your goal is to survive. To survive, you must ad ust. ou must try to maintain emotional control as uickly as possible after the capture. Maintaining emotional control helps you keep control of your mental abilities such as situational awareness, udgment, and descisionmaking skills. Remember, most hostages survive an abduction. After the initial shock of capture wears off, both hostage takers and victims stabili e their emotions and begin to plan for the future. The terrorists may divulge information about themselves, their organi ation, their goals, and ob ectives. They may share their demands and may even discuss roles and responsibilities that the hostages have. The hostages begin making an emotional transition from being a victim to being a survivor. But to survive, you must be alert and cautious. Remember that hostage-takers have used sleepers in their hostage operations. A sleeper is really a terrorist posing as a hostage to inform on the real hostages or draw out security personnel. Be careful who you trust.

## Situational Awareness

If you are blindfolded and gagged during transportation, concentrate on sounds, smells, direction of movement, passage of time, conversations of the hostage-takers, and any information that might be useful. For e ample, you might hear train sounds that might indicate you are near a train station or going by railroad tracks. earing a ship's horn would indicate you are crossing a river or near a body of water. Try and draw a mental map of where you are. If you can hear the hostage-takers, try to determine the language they are speaking, key phrases, goals of the abduction, names, weapons carried, and directions taken, such as make a left at famous landmark . Information collected over time might allow you to guess the possible route and the area where they have taken you. All this information will be very useful if you are released or if you escape while the hostage-takers are still holding other hostages.

## Confrontations

To avoid confrontations with any hostage-taker, you should not carry documents or other sensitive or potentially embarrassing items in your briefcase or on your person. If taken hostage, you must be prepared to e plain telephone numbers, addresses, names, and any other items carried at the time of capture. If asked for identification, show your tourist passport. ispose of items such as your military I and official passport as soon as possible. If you are identified as member of the military, you could be perceived as a threat to the hostage-taker's activities. If interrogated, adopt a simple, tenable position and stick with it. elay identifying yourself as military, but do not lie when asked if you are in the military. Rather, you should attempt to skirt the uestion i.e., if you are a formal school instructor, reply I am a teacher. ou should try to convince your captors that they have kidnapped the wrong person. The terrorists may not be convinced, but don t give up. This delaying effort serves only to ma imi e survival during the initial stages of captivity and reduce the terrorist's apprehension that you might be a threat to their activities. Most casualties among hostages occur during the process of capture and initial internment.

## Defense Mechanisms

As a Marine, it is important to understand what is going on in the minds of hostages. ou might observe what you would consider unusual behavior. This behavior is usually a combination of psychological effects that terrorists seek to achieve by their controlling actions and unconscious, personality-based responses hostages display while in captivity.

Survival is instinctively the most important issue to the human mind. hen placed in a hostage situation, the mind commonly employs defense mechanisms. These unconscious psychological ad ustments are made by hostages to deal with the stress and trauma of the situation. To survive this ordeal, the mind can typically deny that the incident is occurring regress into a dependent state and/or identify with the hostage-taker's demands and values to avoid punishment. A combination of these defense mechanisms can result in the Stockholm Syndrome, whereby the hostage identifies with the hostage-taker and may actively support the hostage-taker's activities.

## Denial

enial is a primitive and very common defense mechanism. To survive an incident that the mind cannot handle, it reacts as if the incident is not happening. ostages commonly respond, This can't be happening to me or This must be a bad dream enial is one stage of coping with an impossible turn of events. These thoughts are actually stress-relieving techni ues. Some hostages deny their situation by sleeping.

As time passes, most hostages gradually accept their situation. They find hope in the thought that their fate is not fi ed, begin to view the situation as temporary, and believe they will be rescued soon.

## Regression

Regression is the return to a more elementary thought pattern commonly found in children. ike a child, a hostage is in a state of e treme dependence and sub ect to fright. Unconsciously, the hostage selects a behavior that was successfully used in childhood. The hostage becomes reliant on the hostage-taker ust as if the hostage-taker was a parent, providing food, shelter, and protection from the outside world. If this thought pattern is firmly in place, hostages may view authorities as a threat to the safety being provided by the hostage-takers.

## Identification

ike regression, identifying with the hostage-taker occurs at the unconscious level. The mind seeks to avoid wrath or punishment by mirroring the behaviors and complying with the demands of the hostage-taker.

## The Stockholm Syndrome

Because you are a potential hostage, you must know and understand the Stockholm Syndrome see page -1 . If taken hostage, you will be able to recogni e if it is happening to other hostages. The Stockholm Syndrome is an automatic, unconscious emotional response to the trauma of becoming a victim. bserved around the world, the Stockholm Syndrome occurs when people are e posed to a high level of stress and cast together with others, not of their choosing, into a new level of adaptation. The result is a positive bond that affects both the hostage and the hostage-taker. The positive emotional bond may develop because of the stress of being in a closed room under siege. This bond unites its victims against the outside world. An attitude of it's us against them seems to develop.

The Stockholm Syndrome produces a variety of responses. Minimal responses consist of victims seeing the event through the eyes of their captor. Those deeply influenced respond by recogni ing the terrorist for his gallant act. Responses have also ranged from hostage apathy to actual participation by the hostages in impeding the efforts of rescue forces and negotiation teams. Another response is losing touch with reality and suffering long-term emotional instability.

No one knows how long the syndrome lasts, but the bond seems to be beyond the control of some hostages. They all share common e periences, including positive contact, sympathy for the human ualities of the hostage-takers, and tolerance.

On 23 August 1973, the quiet, early morning routine of the Credit Bank in Stockholm, Sweden, was destroyed by the sound of a submachine gun. Four hostages were held for 131 hours: three women ranging in ages from 21 to 31 and a 25-year old man. They were held by Jan-Erik Olsson, a thief, burglar, and prison escapee. Olsson kept the hostages in an 11-foot by 4-foot crated bank vault, which they came to share with another criminal and former cell mate of Olsson's, Clark Olofsson. Olofsson joined the group after Olsson demanded his release from prison, and the authorities granted his request. Over time, the hostages began to fear the police more than they feared the robbers. In a phone call to Premier Olof Palme, one of the hostages told the Premier that the robbers were protecting them from the police. After the hostages were released, they began to question why they didn't hate the robbers, why they felt as if Olsson and Olofsson had been the ones to give them their lives back, and that they were emotionally indebted to them for this generosity. For weeks after the incident and while under the care of psychiatrists, some of the hostages experienced severe conflicting emotions of fear that Olsson and Olofsson might escape from jail, yet they also felt no hatred for them. This hostage-taking and its resulting conflicting psychological emotions became knows as the Stockholm Syndrome.

## Positive Contact

ostages may develop positive contact with their abductors if they do not have negative e periences for e ample, beatings and rapes . ositive contact also develops if there has been a negative e perience followed by a positive contact. For e ample, if a hostage was beaten by a cruel guard every time the hostage asked for a drink of water, then a kind guard replaced the cruel guard who gives water freely, typically, the hostage will establish a positive contact with the kind guard.

## Human Qualities of the Hostage-Takers

ostage-takers may talk about their own mental abuse and physical suffering. They want their hostages to see them as victims of circumstance rather than aggressors. Unfortunately, hostages may sympathi e with the hostage-taker and forget that he is the one depriving them of their freedom. nce hostages begin to sympathi e with the hostage-taker, they may actually support the hostage-takers' cause.

## Tolerance

umans have an unconscious limit as to how much we will allow ourselves to be abused or how much we can tolerate. hen we are placed in a survival situation, our acceptable tolerance for abuse usually increases in order for us to survive.

## Coping with Captivity

Coping is a rational mental process used by hostages to deal with and ad ust to the problems of a difficult environment. Unlike defense mechanisms that are mostly unconscious reactions based on personality, coping involves conscious and deliberate thoughts and actions. Coping includes such innovative behaviors as ad usting to living conditions, maintaining dignity and respect, dealing with fear, maintaining mental and physical fitness, and building rapport with captors.

## Living Conditions

The living conditions hostages have endured vary from incident to incident. ostages have been held for days in a bus, airliner or train where heat or lack of heat and lack of ade uate water, food, and toilet facilities were almost unbearable. uring the sei ure of an office or residence, hostages may be in familiar, comfortable surrounding where they have worked or lived. But kidnap victims are fre uently forced to live in makeshift prisons located in attics, basements or remote hideouts. These prisons may be uite small and in some cases prevent the hostage from easily standing or moving around. Sleeping and toilet facilities may be poor, consisting of a cot or mattress and a bucket or tin can for body waste, or a hostage may be forced to soil his living space as well as himself.

The hostage-takers may move you to different holding areas to keep you hidden from authorities. To assist authorities in locating you, you should leave your fingerprints wherever possible in your living area.

## Dignity and Self-Respect

Maintaining one's dignity and self-respect can be very difficult, but it is vital to your survival. our dignity and self-respect may be the keys to retaining your status as a human being in the eyes of the hostage-takers. If you can build empathy while maintaining your dignity, you can potentially lessen the aggression of a captor. Most people cannot inflict pain on another person unless that person becomes dehumani ed or turned into a symbol of their hatred.

## Fear

Fear of death is a hostage-taker's most important tool. They use it to control, intimidate, and wear down the hostage and the negotiators. The fear of death is usually greatest during the first few hours of capture. ostage-takers may induce fear by loading and unloading weapons in the hostage's presence, displaying e cesses of temper, resorting to physical abuse, and staging mock e ecutions which are mercifully stopped at the last minute. As this fear subsides, a hostage may begin to hear he owes his life to the captors who have allowed him to live. Anticipate isolation and terrorist efforts to confuse you. Fear of dying is real, and it can become overwhelming, especially during the early phase of captivity. owever, you must try to maintain emotional control in order to stay mentally alert. Fight despair and depression by keeping a positive mental outlook. Remember, although death is a real possibility, most hostages walk out of the ordeal.

## Physical and Mental Fitness

If abducted, you should develop and maintain a daily physical fitness program. It will help you ward off boredom and can reduce stress. Staying physically fit might be the deciding factor if an escape opportunity presents itself, and you have to run or walk a considerable distance to reach safety. It may be hard to e ercise because of cramped space or physical restraints, but you can run in place or perform isometric e ercises. owever, you should avoid e cessive e ercise that could result in in ury.

It is important to make some mental link to the outside world. To stimulate your mind, you can read, write, daydream, or use your imagination to build something step-by-step a house, a car, a piece of furniture, etc. . Ask the hostage-takers for reading materials or a radio. If possible, communicate with and try to reassure fellow hostages. If it is your day of worship, mentally walk through the various parts of the worship service. stablish a slow, methodical routine for every task.

Typically, hostage-takers want to keep their hostages alive and well. at whatever food is available to maintain your strength. If you need medicine, ask for e actly what you need. If your abductors want you alive, they are not likely to take chances by providing you with the wrong medicine. A side effect of captivity for some hostages is weight loss. Although this may be considerable, it generally does not cause health problems. ostages may also suffer gastrointestinal upset including nausea, vomiting, diarrhea, and/or constipation. Although these symptoms can be debilitating, they are usually not life-threatening.

## Establishing Rapport

Rapport-building techni ues help you make a transition from a faceless symbol who has been dehumani ed to one who is human again. owever, don't e aggerate your human emotions by begging or crying. An emotional outburst could spread panic and fear among the other hostages and could be viewed as a disgusting display of cowardice by the hostage-takers. ou must portray yourself as a person rather than an ob ect by maintaining your dignity, self-respect, and apparent sincerity. ou must attempt to establish rapport with your hostage-taker, but you must do it with dignity and self-respect. This rapport may save your life. ou should

- Make eye contact with the hostage-takers.
- Greet the hostage-takers and use personal names.
- Smile.
- Talk to the hostage-takers. specially talk about your family and show photos if you have them.
- etermine if you have common interests e.g., sports, hygiene, food, etc.
- isten to the hostage-taker. If he wants to talk about his cause, act interested. ou may e plain that you might not agree with him, but you're interested in his point view.
- Avoid appearing overly attentive or interested, the hostage-takers may view this as patroni ing or insincere.
- Avoid arguing with the hostage-takers. Avoid escalating tensions with words such as gun, kill, or punish that could cause the hostage-takers to single you out as being argumentative or combative and therefore a threat to the their authority. Bring up neutral topics at critical times to defuse arguments and reduce tensions.
- Avoid emotionally charged topics of religion, economics, and politics.
- o not refuse favors offered by your captors if doing so will aggravate them or cause further harm to the health and safety of all hostages. owever, do not accept favorable treatment at the e pense of other hostages. Terrorists commonly employ this controlling tactic to cause division and distrust among the hostage group.

## Exploitation of Hostages

ostages should make reasonable efforts to avoid providing oral or signed confessions, answering uestionnaires, making propaganda broadcasts, and conducting news interviews. These actions help terrorist groups further their goals and e ploit the media. Interviews broadcast around the world could embarrass the U.S. or host governments. owever, if you don't comply with the hostage-takers re uests, you could be tortured or threatened with death. ou should never mistake pride for inappropriate resistance. eep your temper under control and maintain a polite bearing. hen being interrogated, take a simple, tenable position and stick with it. Give short answers to the terrorist uestions that discuss unimportant topics. The epartment of efense policy is *survive with honor.* If you are forced to sign or make a statement for the hostage-takers, try to degrade the propaganda, provide minimum information, and avoid making a plea on your behalf. Identify your statement as being made in response to the demands of your captors. o not hide your face if the hostage-takers take photographs of the hostages photos provide authorities with positive identification and information.

## Releases and Rescue Attempts

istorically, the more time that passes, the better chance a hostage has of being released or rescued. The ma ority of hostage-taking incidents are resolved by negotiated releases not rescue attempts. hile the passage of time without rescue or release can be depressing, it is actually to your advantage. Time can produce a positive or negative bond between you and your abductors. If the hostage-taker does not abuse you, hours spent together will most likely build rapport, produce positive results, and increase your chances of survival. owever, you must also look ahead and plan for your release or rescue. ou must also remember that if the hostage-takers' demands are not met, they may kill hostages. ou must prepare yourself for the potential response from authorities if a hostage is killed. Typically, negotiations cease, and rescue forces move in to rescue hostages.

## Releases

The moment of imminent release, like the moment of capture, is very dangerous. The hostage-takers, as well as the hostages, are likely to feel threatened and even panic. The hostage-takers will be e tremely nervous during any release phase, especially if negotiations are drawn out. The terrorists will be an ious to evade capture and punishment, and they will fear being double-crossed by the authorities. ou need to pay close attention to the instructions the hostage-takers give you when the release takes place: do not panic and do not run because the hostage-takers may shoot you.

## Rescue Attempts

uring the rescue attempt, both the hostage and the rescue force are in e treme danger. Most hostages who die are killed during rescue attempts. ou must be especially alert, cautious, and obedient to instructions if an attempt is imminent or is occurring. If possible, position yourself in the safest area, such as under desks, behind chairs, or behind any large ob ect that provides protection. ou should avoid being near doors, windows, or open areas. If the doors fly open followed by rescue forces, drop to the floor immediately, lie as flat as possible, do not move, do not say anything, do not attempt to pick up a weapon or help the rescuers.

Rescue forces have no idea whether you are friend or foe. Any movement you make could result in in ury or death to you or your fellow hostages. It could also distract members of the rescue force, which, in turn, could lead to in uries or deaths among the rescuers. uring a rescue operation at ntebbe, Uganda, a woman hostage threw her hands up in a natural gesture of oy as the rescue forces came bursting in. Unfortunately, the rescue forces shot her. nce the rescue forces are in control, you might be handled roughly and ordered up against the wall. ou will probably be handcuffed, searched, and possibly gagged and/or blindfolded until everyone is positively identified.

## After the Release

nce you are safely in the hands of the authorities, remember to cooperate fully with them, especially if others are still being held. As soon as you can, write down everything you can remember: guard location, weapons and e plosives description and placement, and any other information that might help rescue forces.

After your release, you must prepare yourself for the aftermath. The news media will want an interview immediately, and you will be in no condition to provide intelligent, accurate responses. o not make comments to the news media until you have been debriefed by proper U.S. authorities and have been cleared to do so by the appropriate military commanders. ou should only say that you are grateful to be alive and thankful for being released. ou should not say anything that could harm fellow hostages who may still be in captivity. ou must not say anything that is sympathetic to the terrorist cause or that might gain support for them.

Upon release, many hostages feel guilty for not having conducted themselves in a heroic manner. motional turmoil is common.

Some may feel angry because they feel that their government did not do enough to protect them. Remember that a government's unwillingness to make concessions to terrorists discourages future acts of terrorism and sends a message to all terrorists worldwide. hen ransoms for captives of terrorists have been paid by governments, these payments have usually been used by terrorists to increase their status and capability to continue terrorist acts. It did not mean that your life had no value. It is the policy of the United States that when Americans are abducted overseas, the United States will continue to cultivate international cooperation to combat terrorism and secure the safe release of the hostages.

# Appendix A
# Vehicle Bomb Search

## Prevention

 ou must learn to recogni e danger signs and how to search a vehicle for tampering or e plosive devices. Certain procedures apply that may help prevent you from becoming a victim of a vehicle bombing, they include

- Checking your vehicle at irregular times to prevent establishing a pattern.
- Being suspicious and aware of what is going on around you.

YOU CAN PREVENT THIS

- ocking your vehicle and parking in secured areas whenever possible to limit easy access.

- Allowing a fine coat of dust to remain on the vehicle surface or applying talcum powder.

- Securing transparent tape to vehicle doors, trunk, and hood to help detect tampering.

- Installing two bolts in an   pattern over the open end of the e haust pipe.

- Installing a locking gas cap and a mesh strainer in the mouth of the filler tube.

- Getting out of the car to wait for passengers.

BE SUSPICIOUS

## Exterior Search

Unless your vehicle has been under 24-hour guard, you must always assume that it has been rigged with a bomb, and you must use e treme caution while conducting an e ternal search. ou must know your vehicle inside and out so that you can uickly recogni e something that is wrong. If possible, you should search in pairs. If you notice that you are being observed while conducting the search, gently close the hood, trunk, or door if it is open and walk away from the vehicle. If you find a bomb, or something that looks like a bomb, N T T UC IT: immediately contact an e plosive ordnance disposal unit. An e terior search is conducted as follows:

- Search the area around the vehicle. Bushes, shrubs, trees, trash cans, mailbo es, etc., are areas where an improvised e plosive device may be placed or concealed. An e plosion close to a vehicle can produce the same devastating effects as if the bomb was placed in the vehicle.

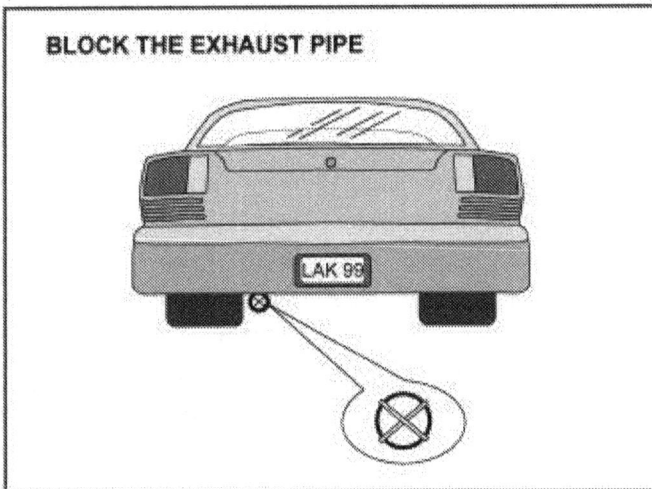

BLOCK THE EXHAUST PIPE

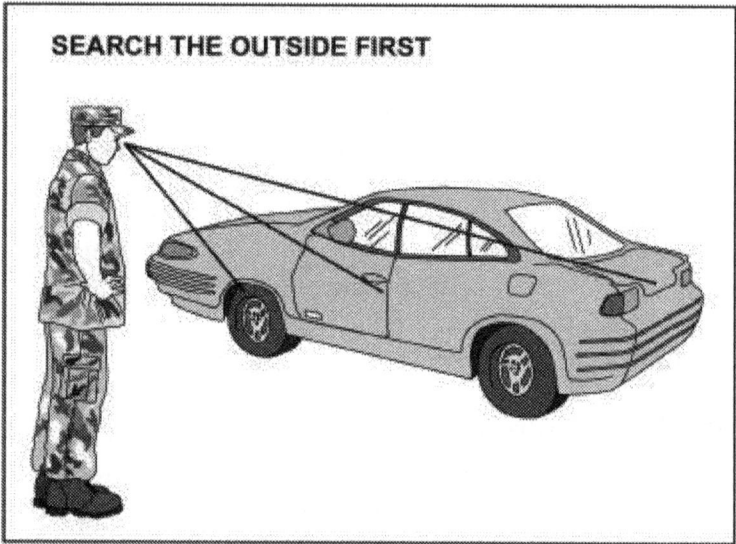

SEARCH THE OUTSIDE FIRST

- amine the e terior surfaces of the vehicle. ook for signs of tampering  wires hanging down  or doors, hood, or trunk left a ar. oes anything appear different from when you left the vehicle

- amine the film of dust or talcum powder. Is it undisturbed  as another layer of dust appeared

- amine the transparent tape  removed or broken  .

- amine the hood or trunk lock  look, do not lift .  as either been  immied

- amine the vehicle for other signs of forced entry  broken windows, scratched paint, bent, or damaged metal .

- pen the interior compartments  ust a little, enough to gently run your fingers around the opening and to feel for a trip wire.

- If a trip wire is not found, you can open the compartment gradually, e amining the hinges for pressure or tension release initiators.

-  ook closely for any bits of tape or wire lying in or around the vehicle.

-  amine the ground for any unusual marks or signs of digging on the ground.

- Remove the gas cap and look inside.

- Check in and around the e haust pipe.

- Check the undercarriage. If you can, use a long-handled mirror to help in your search.

CHECK WITH A MIRROR

- amine the wheel wells and behind the bumpers.
- amine everywhere: steps, handholds, even the canvas tops of tactical vehicles.

## Interior Search

ou must conduct an interior search with e treme caution. Always *look* inside before you *move* inside. ou must avoid touching anything in the interior of the vehicle until it has been searched, and you should never rest your hand on the seat. If you find a bomb, or something that looks like a bomb,    N T T UC  IT: immediately contact an e plosive ordnance disposal unit. An interior search is conducted as follows:

- ook through the windows. o you see anything out of place  as anything been moved  as anything been added: a package or briefcase that does not belong there  o you see tapes or wires hanging down
- Unlock a door, open it very slowly, and only a uarter of an inch. ook around the door edges for trip wires. If the door looks free of trip wires, open the door gradually, e amining the hinges for pressure or tension release initiators. If anything looks suspicious, close the door gently.
- ook at the carpet or floor mats for any suspicious bulges. ook as far as you can under the seat, around the seat, and behind the seat without entering the vehicle.
- ook around and behind all the other seats.

- Slip into the seat and check the ashtray, ad ustable headrest, and seat belt.

- amine the right rear passenger seat.

- Check the glove compartment.

- ook under the dash, checking especially for strange tapes and wires.

- Use a flashlight to check the air conditioning ducts and other vehicle cavities.

- amine the sun visors and mirrors for signs of tampering.

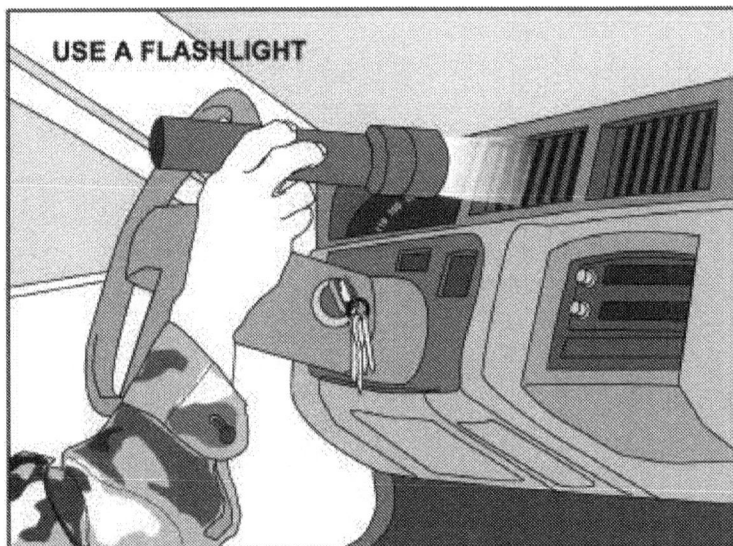

USE A FLASHLIGHT

# Engine and Trunk Search

nce an interior search has been conducted, you should e it the vehicle and open the engine hood and trunk. ou should open the hood or the trunk only a uarter of an inch at first and very gently feel for wires along the length of the hood or the trunk.

An engine or trunk search is conducted as follows:

## Engine

- Raise the hood and make a thorough search of the engine compartment and fire wall.

- ook for any strange or new-looking wires attached to the battery, clutch, coil, accelerator, or any power-operated e uipment.

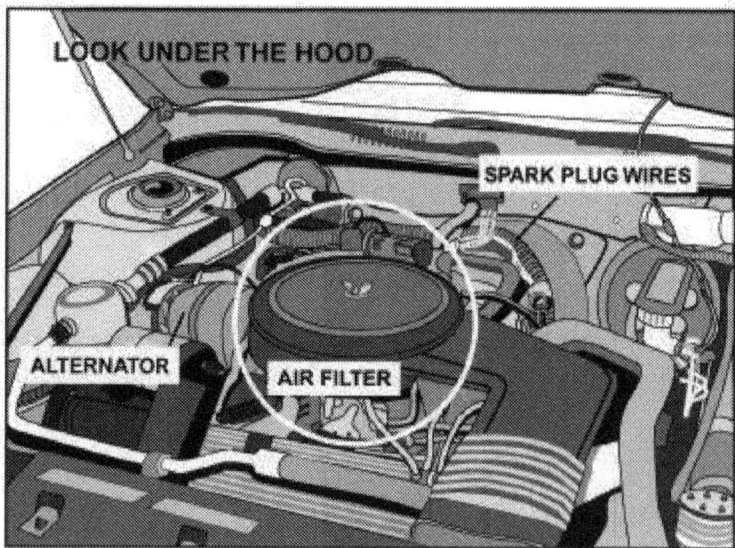

- Check engine cavities for anything that looks like it does not belong there and anything out of place.

- pen the air filter and look inside. ay special attention to the spark plug wires, the distributor, the ignition area, and the e haust manifold.

## Trunk

- Raise the trunk and make a thorough search.

- Check the items inside. Is anything rearranged  Is anything new

- Check the spare tire to ensure it is filled with air.

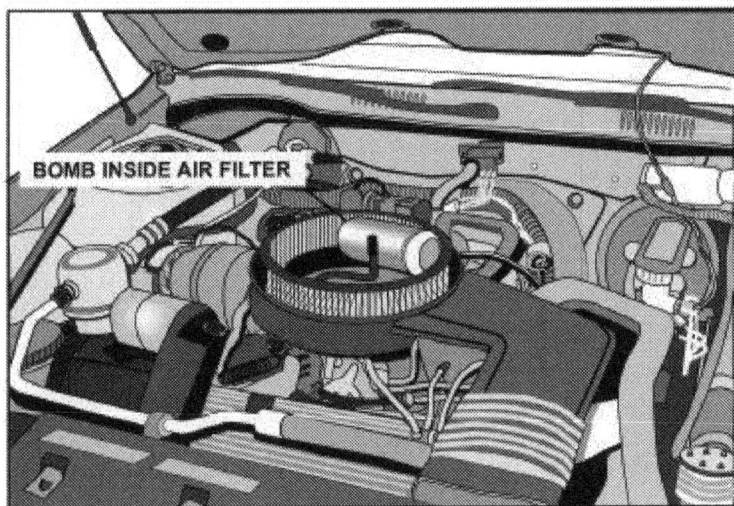

BOMB INSIDE AIR FILTER

## Final Check Before Starting the Vehicle

 nce you have inspected the e terior and interior of the vehicle, the trunk, and the engine, you can get into the driver's seat and check the dashboard. ou should look at the turn signals and lights, if nothing is unusual, then turn them on. ou also need to check and test the wipers, washer, radio, and horn. nce you feel that the vehicle is safe, you can start the vehicle. et the vehicle run for about 2 minutes before you proceed to your destination.

# Appendix B
# Joint Staff Guide 5260,
# "Service Member's
# Personal Protection Guide:
# A Self-Help Guide to Combat Terrorism
# while Overseas"

## FOREWORD

This guide is designed to assist in making you and your family less vulnerable to terrorists while stationed or traveling overseas. ou should become familiar with its contents and incorporate those protective measures that are applicable to your particular situation. Moreover, ensure every member of your family is made aware of this valuable information so they can help protect themselves as well.

Terrorism is an indiscriminate crime that strikes in varying forms of threats and violence. Terrorist generate fear through intimidation, coercion, and acts of violence such as hi ackings, bombings or kidnappings, which usually occur more fre uently in certain parts of the world, making travelers to foreign countries more likely potential victims. As past events have shown, terrorists have reached new levels of organi ation, sophistication, and violence their tactics and techni ues are always changing and will continue to be a challenge to neutrali e. Accordingly, we must remain diligent in applying the proper protective measures.

ou and your family are an important part our military. This guide will not ensure immunity from terrorism, but by practicing these techni ues and proven security habits, the possibility of becoming a target will be reduced. efensive awareness and personal security regarding terrorism are responsibilities of everyone assigned to  . As members of the military community, you are highly valuable yet most valuable resource. Constant awareness can help protect all members of the military family from acts of terrorism.

NR  .S  T N
Chairman
of the  oint Chiefs of Staff

# Joint Staff Guide 5260

# Table of Contents

# STEPS TO COMBAT TERRORISM

## Keep a Low Profile

our dress, conduct, and mannerisms should not attract attention. Make an effort to blend into the local environment. Avoid publicity and don t go out in large groups. Stay away from civil disturbances and demonstrations.


## Be Unpredictable

ary your route to and from work and the time you leave and return home. ary the way you dress. on t e ercise at the same time and place each day, never alone, on deserted streets, or country roads. et people close to you know where you are going what you ll be doing, and when you should be back.

## Be Alert

atch for anything suspicious or out of place. on t give personal information over the telephone. If you think you are being followed, go to a preselected secure area. Immediately report the incident to the military/security police or law enforcement agencies. In overseas areas without such above agencies report the incident to the Security fficer or the Military Attache at the U.S. mbassy.

# Section I. General Security Checklist

- Instruct your family and associates not to provide strangers with information about you or your family.

- Avoid giving unnecessary personal details to anyone.

- Be alert to strangers who are on government property for no apparent reason. Report all suspicious persons loitering near your office  attempt to provide a complete description of the person and/or vehicle to policy or security.

- ary daily routines, such as departure times and routes to and from work, to avoid habitual patterns.

- Refuse to meet with strangers outside your work place.

- Always advise associates or family members of your destination and the anticipated time of arrival when leaving the office or home.

- on t open doors to strangers.

- Memori e key phone numbers office, home, police, security, etc.

- Be cautious about giving out information regarding family travel plans or security measures and procedures.

- earn and practice a few key phrases in the native language, such as  I need a policeman, doctor,  etc.

## House, Home, and Family Security

Although spouses and children are seldom targeted by terrorists, they should practice basic precautions for their personal
security. Familiari e your family with the local terrorist threat
and regularly review the protective measures and techni ues
listed in this handbook. nsure everyone in the family knows
what to do in an emergency.

### Tips for the Family at Home

- Restrict the possession of house keys. Change locks if keys
  are lost or stolen and when moving into a previously occupied
  residence.

- ock all entrances at night, including the garage. eep the
  house locked, even if you are at home.

- estroy all envelopes or other items that indicated your name
  and rank.

- evelop friendly relations with your neighbors.

- on't draw attention to yourself. Be considerate of neighbors.

- Avoid fre uent e posure on balconies and near windows.

### Be Suspicious

- Be alert to public works crews and other foreign nationals
  re uesting access to residence  check their identities through a
  peep-hole before allowing entry.

- Be cautious about peddlers and strangers.

- rite down license numbers of suspicious vehicles note descriptions of occupants.

- Treat with suspicion any in uiries from strangers concerning the whereabouts or activities of family members.

- Report all suspicious activity to Military/Security olice or local law enforcement.

## Telephone Security

- ost emergency numbers on the telephone and preprogram phone numbers where possible:

  Military/Security olice:
  ocal olice:
  Fire epartment:
  ospital:
  Ambulance:

- o not answer your telephone with your name and rank.

- Report all threatening phone calls to security officials and telephone company.

## When Going Out Overseas

- Travel in small groups as much as possible. Avoid high risk areas such as demonstrations, and vary movements so as not to be predictable.

- Try to be inconspicuous when using public transportation and facilities. ress, conduct, and mannerisms should not attract attention.

- o not be curious about spontaneous gatherings or demonstrations. Avoid them.

- Stay away from known trouble or disreputable places  visit only reputable establishments, but don t fre uent the same off-base locations  in particular, known, U.S.-associated locales .

- now emergency numbers and how to use the local telephone system.

## Special Precautions for Children

- now where your children are all the time.

- Never leave young children alone or unattended. Be certain they are in the care of a trustworthy person.

- If it is necessary to leave children at home, keep the house will lighted and notify the neighbors.

- Instruct children to keep doors and windows locked, and to never admit strangers.

- Teach children how to contact the police or a neighbor in an emergency.

- Advise your children to:

  - Never leave home without telling you where they will be and who will accompany them.

  - Travel in pairs or groups.

  - Avoid isolated areas.

  - Use locally approved play areas where recreational activities are supervised by responsible adults and where police protection is readily available.

  - Refuse automobile rides from strangers and refuse to accompany strangers anywhere on foot even if the strangers say mom or dad sent them, or said it was  okay.

- Report immediately to the nearest person of authority parent, teacher, police anyone who attempts to molest or annoy them.

## Security Precautions When You're Away

- eave the house with a lived-in look.

- Stop deliveries of or forward mail to a neighbor s home.

- on t leave notes on doors.

- on't hide keys outside house.

- Use a timer appropriate to local electricity to turn lights on and off at varying times and locations.

- eave radio on.

- ide valuables.

- Notify the police or trusted neighbor of your absence.

- Ask a trusted friend/neighbor to periodically check residence.

## Suspicious Packages or Mail

- Suspicious characteristics to look for include:

  - An unusual or unknown place of origin.

  - No return address.

  - An e cessive amount of postage.

  - Abnormal or unusual si e.

  - ily stains on the package.

  - ires or strings protruding from or attached to an item.

  - Incorrect spelling on the package label.

- iffering return address and postmark.

- Appearance of foreign style handwriting.

- eculiar odor. Many e plosive used by terrorists smell like shoe polish or almonds.

- Unusual heaviness or lightness.

- Uneven balance or shape.

- Springiness in the top, bottom, or sides.

- Never cut tape, strings or other wrappings on a suspect package or immerse a suspected letter or package in water. ither action could cause an e plosive devise to detonate.

- Never touch or move a suspicious package or letter.

- Report any suspicious packages or mail to security officials immediately.

## Domestic Employees

- Conduct a security background check with local police, neighbors, and friends.

- Inform employees about security responsibilities.

- Instruct them which phone or other means of communication to use in an emergency.

- o not discuss travel plans or sensitive topics within earshot of domestic employees who have no need to know.

- iscuss duties in friendly, firm manner.

- Give presents or gratuities according to local customs.

## Residential Security

- terior grounds:

  - o not put your name on the outside of your residence or mailbo .

  - ave good lighting.

  - Control vegetation to eliminate hiding places.

- ntrances and e its should have:

  - Solid doors with deadbolt locks.

  - ne-way peep-holes in door.

  - Bars and locks on skylights.

  - Metal grating on glass doors and ground floor windows, with interior release mechanisms that are not reachable from outside.

- Interior features:

  - Alarm and intercom systems.

  - Fire e tinguishers.

  - Medical and first aid e uipment.

- ther desirable features:

  - A clear view of approaches.

  - More than one access road.

  - ff-street parking.

  - igh - feet perimeter wall or fence.

## Ground Transportation Security

Criminal and terrorist acts against individuals usually occur outside the home and after the victim s habits have been established. our most predictable habit is the route of travel from home to duty station or to commonly fre uented local facilities.

### Vehicle Overseas

- Select a plain car avoid the rich American look.
- Consider not using a government car that announces ownership.
- o not display decals with military or unit affiliations on vehicle.
- o not openly display military e uipment or field gear in your vehicle.

### Auto Maintenance

- eep vehicle in good repair.
- Always keep gas tank at least half full.
- nsure tires have sufficient tread.

### Parking Your Car

- Always lock your car.
- on t leave your car on the street overnight, if possible.
- Never get out without checking for suspicious persons. If in doubt, drive away.
- eave only the ignition key with parking attendant.

- on t leave garage doors open or unlocked.

- Use a remote garage door opener if available. nter and e it your car in the security of the closed garage.

## On the Road

- Before leaving buildings to get into your vehicle, check the surrounding area to determine if anything of a suspicious nature e ists. isplay the same wariness before e iting your vehicle.

- rior to getting into a vehicle, check beneath it for any tampering or bombs by looking for wires, tape, or anything unusual.

- If possible, vary routes to work and home.

- Avoid late night travel.

- Travel with companions.

- Avoid isolated roads or dark alleys when possible.

- abitually ride with seatbelts buckled, doors locked, and windows closed.

- o not allow your vehicle to be bo ed in  maintain a minimum  -foot interval between you and the vehicle in front  avoid the inner lanes. Be alert while driving or riding.

- now how to react if you are being followed:

  - Check during turns for confirmation of surveillance.

  - o not stop or take other actions which could lead to confrontation.

  - o not drive home. If necessary, go to the nearest military base or police station.

  - Get description of car and its occupants.

  - Report incident to military/security police.

- Recogni e events that signal the start of an attack. hen one of these events occurs, start mentally preparing a course of action in case an attack develops. These events may include, but are not limited to:

  - Cyclist falling in front of your car.

  - Flagman or workman stopping your car.

  - Unusual or false police or government checkpoint.

  - Fake police or government checkpoint.

  - isabled vehicle/accident victims on the road.

  - Unusual detours.

  - An accident in which your car is struck.

  - Cars or pedestrian traffic that bo  you in.

  - Sudden activity or gunfire.

- now what to do if under attack in a vehicle:

  - ithout sub ecting yourself, passengers, or pedestrians to harm, try to draw attention to your car by sounding the horn.

  - ut another vehicle between you and your pursuer.

  - ecute immediate turn and escape  ump the curb at  0- 4  degree angle,   mph ma imum.

  - Ram blocking vehicle if necessary.

  - Go to closest safehaven.

  - Report incident to military/security police.

## Commercial Buses, Trains, and Taxis

- ary mode of commercial transportation.
- Select busy stops.
- o not always use the same ta i company.
- o not let someone you don t know direct you to a specific cab.
- nsure ta i is licensed, and has safety e uipment  seatbelts at a minimum .
- nsure face of driver and picture on license are the same.
- Try to travel with a companion.
- If possible, specify the route you want the ta i to follow.

## Traveling Defensively by Air

Air travel, particularly through high risk airports or countries, poses security problems different from those of ground transportation. ere, too, simple precautions can reduce the ha ards of a terrorist assault.

### Making Travel Arrangements

- Get a threat briefing from your security officer, antiterrorism training officer, or force protection officer prior to traveling in a high risk area.  our force protection officer will know which area    considers a high risk area.

- Before traveling, consult the    Foreign Clearance Guide to ensure you know and can meet all re uirements for travel to a particular country.

- Use military air or U.S. flag carriers.

- Avoid scheduling through high risk areas  if necessary, use foreign flag airlines and/or indirect routings to avoid high risk airports.

- o not use rank or military address on tickets, travel documents, or hotel reservations.

- Select a window seat, which would offer more protection since aisle seats are closer to the hi ackers  movements up and down the aisle.

- Rear seats also offer more protection since they are farther from the center of hostile action which is often near the cockpit.

- Seats at an emergency e it may provide an opportunity to escape.

- Avoid off-base hotels  use government  uarters or contracted hotels.

## Personal Identification

- o not discuss your military affiliation with anyone.

- ou must have proper identification to show airline and immigration officials. Consider use of a tourist passport, if you have one with necessary visas, providing the country you are visiting allows it.

- If you use a tourist passport, consider placing your official passport, military I , travel orders, and related documents in your checked luggage, not in your wallet or briefcase.

- If you must carry these documents on your person, select a hiding place onboard the aircraft to  ditch  them in case of a hi acking.

- o not carry classified documents unless they are absolutely mission-essential.

## Luggage

- Use plain, civilian luggage avoid military-looking bags such as B-4 bags and duffel bags.

- Remove all military patches, logos, or decals from your luggage and briefcase.

- nsure luggage tags don t show your rank or military address.

- o not carry official papers in your briefcase.

## Clothing

- Travel in conservative civilian clothing when using commercial transportation or when traveling military airlift if you are to connect with a flight at a commercial terminal in a high risk area.

- on't wear distinct military items such as organi ational shirts, caps, or military issue shoes or glasses.

- on't wear U.S.-identified items such as cowboy hats or boots, baseball caps, American logo T-shirts, ackets, or sweatshirts.

- ear a long-sleeved shirt if you have a visible U.S.-affiliated tattoo.

## Precautions at the Airport

- Arrive early watch for suspicious activity.

- ook for nervous passengers who maintain eye contact with others from a distance. bserve what people are carrying. Note behavior not consistent with that of others in the area.

- No matter where you are in the terminal, identify ob ects suitable for cover in the event of attack pillars, trash cans, luggage, large planters, counters, and furniture can provide protection.

- roceed through security checkpoints as soon as possible.

- Avoid secluded areas that provide concealment for attackers.

- Be aware of unattended baggage anywhere in the terminal.

- Be e tremely observant of personal carry-on luggage. Thefts of briefcases designed for laptop computers are increasing at airports worldwide. ikewise, luggage not properly guarded provides an opportunity for a terrorist to place an unwanted ob ect or device in your carry-on bag. As much as possible, do not pack anything your cannot afford to lose if the documents are important, make a copy and carry the copy.

- bserve the baggage claim area from a distance. o not retrieve your bags until the crowd clears. roceed to the customs lines at the edge of the crowd.

- Report suspicious activity to the airport security personnel.

## Actions if Attacked in an Airport

- ive for cover. o not run. Running increases the probability of shrapnel hitting vital organs or the head.

- If you must move, belly crawl or roll. Stay low to the ground, using available cover.

- If you see grenades, seek immediate cover, lay flat on the floor, feet and knees tightly together with soles toward the grenade. In this position, your shoes, feet, and legs protect the rest of your body. Shrapnel will rise in a cone from the point of detonation, passing over your body.

- lace arms and elbows ne t to your ribcage to protect your lungs, heart, and chest. Cover your ears and head with your hands to protect neck, arteries, ears, and skull.

- Responding security personnel will not be able to distinguish you from attackers. o not attempt to assist them in any way. ay still until told to get up.

## Actions if Hijacked

- Remain calm, be polite and cooperate with your captors.

- Be aware that all hi ackers may not reveal themselves at the same time. A lone hi acker may be used to draw out security personnel for neutrali ation by other hi ackers.

- Surrender your tourist passport in response to a general demand for identification.

- on t offer any information  confirm your military status if directly confronted with the fact. Be prepared to e plain that you always travel on your personal passport and that no deceit was intended.

- iscretely dispose of any military or U.S.-affiliated documents.

- on't draw attention to yourself with sudden body movements, verbal remarks, or hostile looks.

- repare yourself for possible verbal and physical abuse, and lack of food, drink, and sanitary conditions.

- If permitted, read, sleep, or write to occupy your time.

- iscretely observe your captors and memori e their physical descriptions. Include voice patterns and language distinctions, as well as clothing and uni ue physical characteristics.

- Cooperate with any rescue attempt. ie on the floor until told to rise.

# Taken Hostage–You Can Survive!

The chances of you being taken hostage are truly remote. ven better news is that survival rates are high. But should it happen, remember, your personal conduct can influence treatment in captivity. The epartment of State has responsibility for all U.S. government personnel and their dependents in overseas areas. Should a hostage situation develop, the epartment of State will immediately begin to take action according to preconceived plans to attempt to release the hostages. If kidnapped and taken hostage, the hostage has three very important rules to follow:

1. **Analyze the problem so as not to aggravate the situation.**

2. **Make decisions to keep the situation from worsening.**

3. **Maintain discipline to remain on the best terms with the captors.**

## Preparing the Family

- ave your family affairs in order, including an up-to-date will, appropriate powers of attorney, and measures taken to ensure family financial security.

- Issues such as continuing the children s education, family relocation, and disposition of property should be discussed with family members.

- our family should know that talking about your military affiliation to non- people may place you, or them, in great danger.

- They must be convinced the U.S. government will work to obtain your safe release.

- on't be depressed if negotiation efforts appear to be taking a long time. Remember, your chances of survival actually increase with time.

## Stay in Control

- Regain your composure as soon as possible and recogni e your fear. our captors are probably as apprehensive as you, so your actions are important.

- Take mental notes of directions, times of transit, noises, and other factors to identify your location.

- Note the number, physical description, accents, habits, and rank structure of your captors.

- Anticipate isolation and efforts to disorient and confuse you.

- To the e tent possible, try to mentally prepare yourself for the situation ahead. Stay mentally active.

## Dealing with Your Captors

- o not aggravate them.

- o not get into political or ideological discussions.

- Comply with instructions, but always maintain your dignity.

- Attempt to develop a positive relationship with them.

- Be proud of your heritage, government, and military association, but use discretion.

## Keep Occupied

- ercise daily.

- Read anything and everything.

- at what is offered to you. ou must maintain your strength.

- stablish a slow, methodical routine for every task.

## Being Interrogated

- If you need to make up a story to protect sensitive information, take a simple, tenable position you will remember and stick to it.

- Be polite and keep your temper.

- Give short answers. Talk freely about nonessential matters, but be guarded when conversations turn to matters of substance.

- o not be lulled by a friendly approach. Remember, one terrorist may play Good Guy and one Bad Guy. This is the most common interrogation techni ue.

- Briefly affirm your belief in basic democratic principles.

- If forced to present terrorist demands to authorities, in writing or on tape, state clearly that the demands are from your captors.

- Avoid making a plea on your behalf.

## During Rescue

- rop to the floor and be still. Avoid sudden moves. ait for instruction.

- nce released, avoid derogatory comments about your captors such remarks will only make things harder for those still held captive.

## Responding to Chemical Threats

Chemical agents are generally li uids, often aerosoli ed, and although some effects are delayed, most induce an immediate response. There are many different potential chemical agents that a terrorist could use as a weapon. Nonetheless, the following broad generali ations can be made:

- Although food or water contamination is possible, inhalation is the most likely method of delivery. rotection of the breathing airway is the single most import factor of defense.

- Many likely agents are heavier than air and will tend to stay close to the ground. This dictates an upward safety area strategy.

- Generally, chemical agents tend to present an immediate noticeable effect. Medical attention should be sought immediately, even if e posure is thought to be limited.

- Most chemical agents that present an inhalation ha ard will break down fairly rapidly when e posed to sun, diluted with water, or dissipated in high winds.

- No matter what the agent or particular concentration, evacuation preferable upwind from the area of attack is always advisable unless you are properly e uipped with appropriate breathing device and protective clothing.

## Detection

A chemical attack or incident will not always be immediately apparent because many agents are orderless and colorless. Be alert to the possible presence of an agent. Indicators of such an attack includes:

- roplets of oily film on surfaces.

- Unusual dead or dying animals in the area.

- Unusual li uid sprays or vapors.

- Une plained odors  smell of bitter almonds, peach kernels, newly mowed hay, or green grass .

- Unusual or unauthori ed spraying in the area.

-  ow-lying clouds of fog unrelated to weather, clouds of dust or suspended, possible colored particles.

-  eople dressed unusually  long-sleeved shirts or overcoats in the summertime  or wearing breathing protection particularly in areas where large numbers of people tend to congregate, such as subways, or stadiums.

-  ictims displaying symptoms of nausea, difficulty breathing, convulsions, disorientation, or patterns of illness inconsistent with natural disease.

## Defense in Case of Chemical Attack

rotection of breathing airways is the single most important thing a person can do in the event of chemical attack. In most cases, absent a gas mask, the only sure way to protect an airway is to put distance between you and the source of the agent. hile evacuating the area, cover your mouth and nose with a handkerchief, coast sleeve, or any piece of cloth to provide some moderate means of protection. ther steps are:

- Stay alert. arly detection enhances survival.

- Move upwind from the source of attack.

- If evacuation from the immediate area is impossible, move out-doors or to an interior room on a higher floor. Remember, many agents are heavier than air and will tend to stay close to the ground.

- If indoors and no escape outside is possible, close all windows and e terior doors while also shutting down the air condition-ing or heating systems to prevent circulation of air.

- Cover your mouth and nose. If gas masks are not available, use a surgical mask or handkerchief. An improvised mask can be made by soaking a clean cloth in a solution of one table-spoon of baking soda in a cup of water. Not highly effective, it may provide some protection.

- Cover bare arms and legs and make sure any cuts or abrasions are covered and bandaged.

- If splashed with an agent, immediately wipe it off using copious amounts of warm soapy water or a diluted 10:1 bleach solution.

- If water is not available, talcum powder or flour are also e cellent means of decontamination of li uid agents. Sprinkle the flour or powder liberally over the affected skin area, wait 0 seconds, and gently wipe off with a rag or gau e pad.

No matter what the agent or concentration, medical attention should be sought immediately, even if the exposure is thought to be limited.

# Section II. DOD Policy Guidance on the Code of Conduct for Personnel Subject to Terrorist Activity

## Department of Defense

### DERIVED FROM DOD DIRECTIVE 1300.7 AND DOD INSTRUCTION 1300.21

**A. Policy:** This policy concerning the conduct of U.S. military personnel isolated from U.S. control applies at all times. U.S. military personnel finding themselves isolated from U.S. control are re uired to do everything in their power to follow policy. The     policy in this situation is to survive with honor.

**B. Scope:** The Code of Conduct is a moral guide designed to assist military personnel in combat or being held prisoners of war to live up to the ideals contained in the     policy. This guidance shall assist U.S. military personnel who find themselves isolated from U.S. control in peacetime, or in a situation not related specifically in the Code of Conduct.

**C. Rationale:** U.S. military personnel, because of their wide range of activities, are sub ect to peacetime detention by unfriendly governments or captivity by terrorist groups. This guidance seeks to help U.S. military personnel survive these situations with honor and does not constitute a means for udgment or replace the UCM  as a vehicle for enforcement of proper conduct. This guidance, although not e actly the same as the Code of Conduct, in some areas, applies only during peacetime. The term peacetime means that armed conflict does not e ist or armed conflict does e ist, but the United States is not involved directly.

**D. General:** U.S. military personnel captured or detained by hostile foreign governments or terrorists are often held for purposes of e ploitation of the detainees or captives, or the U.S. Government, or all of them. This e ploitation can take many forms, but each form of e ploitation is designed to assist the foreign government or the terrorist captors. In the past, detainees have been e ploited for information and propaganda efforts, confessions to crimes never committed, all of which assisted or lent credibility to the detainers. Governments also have been e ploited in such situation to make damaging statements about themselves or to force them to appear weak in relation to other governments. Ransoms for captives of terrorists have been paid by government and such payments have improved terrorist finances, supplies, status and operations, often prolonging the terror carried on by such groups.

**E. Responsibility:** U.S. military personnel, whether detainees or captives, can be assured that the U.S. Government will make every good faith effort to obtain their earliest release. Faith in one s country and its way of life, faith in fellow detainees or captives, and faith in one s self are critical to surviving with honor and resisting e ploitation. Resisting e ploitation and having faith in these areas are the responsibility of all Americans. n the other hand, the destruction of such faith must be the assumed goal of all captors determined to ma imi e their gains from a detention or captive situation.

**F. Goal:** very reasonable stop must be taken by U.S. military personnel to prevent e ploitation of themselves and the U.S. Government. If e ploitation cannot be prevented completely, every step must be taken to limit e ploitation as much as possible. In a sense, detained U.S. military personnel often are catalysts for their own release, based upon their ability to become unattractive sources or e ploitation. That is, one that resists successfully may e pect detainers to lose interest in further e ploitation attempts. etainees or captives very often must make their own udgments as to which actions will increase their chances of returning home with honor and dignity. ithout e ception, the military member who can say honestly that he/she has done his/her utmost in a detention or captive situation to resist e ploitation upholds    policy, the founding principles of the U.S., and the highest traditions of military service.

**G. Military Bearing and Courtesy:** Regardless of the type of detention or captivity, or harshness of treatment, U.S. military personnel will maintain their military bearing. They should make every effort to remain calm and courteous, and pro ect personal dignity. This is particularly important during the process of capture and the early stages of internment when the captor may be uncertain of his control of the captives.

**H. Classified Information:** There are no circumstances in which a detainee or captive should voluntarily give classified information or materials to those unauthori ed to receive them. To the utmost of their ability, U.S. military personnel held as detainees, captives, or hostages will protect all classified information. An unauthori ed disclosure of classified information, for whatever reason, does not  ustify further disclosures. etainees, captives and hostages must resist, to the utmost of their ability, each and every attempt by their captor to obtain such information.

**I. Chain of Command:** In group detention, captivity, or hostage situations military detainees, captives or hostages will organi e, to the fullest e tent possible, in a military manner under the senior military member present and eligible to command. The importance of such organi ation cannot be over emphasi ed. istorically, in both peacetime and wartime, establishment of a military chain of command has been a tremendous source of strength for all captives. very effort will be made to establish and sustain communications with other detainees, captives, or hostages. Military detainees, captives, or hostages will encourage civilians being held with them to participate in the military organi ation and accept the authority of the senior military member. In some circumstances, such as embassy duty, military members may be under the direction of a senior U.S. civilian official. Notwithstanding such circumstances, the senior military member still is obligated to establish, as an entity, a military organi ation and to ensure that the guidelines in support of the     policy to survive with honor are not compromised.

**J. Guidance for Detention by Governments:**  nce in the custody of a hostile government, regardless of the circumstances that preceded the detention situation, detainees are sub ect to the laws of that government. In light of this, detainees will maintain military bearing and should avoid any aggressive, combative, or illegal behavior. The latter could complicate their situation, their legal status, and any efforts to negotiate a rapid release.

1. As American citi ens, detainees should be allowed to be placed in contact with U.S. or friendly embassy personnel. Thus, detainees should ask immediately and continually to see U.S. mbassy personnel or a representative of an allied or neutral government.

2. U.S. military personnel who become lost or isolated in a hostile foreign country during peacetime will not act as combatants during evasion attempts. Since a state of armed conflict does not e ist, there is no protection afforded under the Geneva Convention. The civil laws of that country apply. owever, delays in contacting local authorities can be caused by in uries affecting the military member s mobility, disorientation, fear of captivity, or desire to see if a rescue attempt could be made.

. Since the detainer s goals may be ma imum political e ploitation, U.S. military personnel who are detained must be e tremely cautious of their captors in everything they say and do. In addition to asking for a U.S. representative, detainees should provide name, rank, social security number, date of birth, and the innocent circumstances leading to their detention. Further discussions should be limited to and revolve around health and welfare matters, conditions of their fellow detainees, and going home.

4. istorically, the detainers have attempted to engage military captives in what may be called a  battle of wits  about seemingly innocent and useless topics as well as provocative issues. To engage any detainer in such useless, if not dangerous, dialog only enables a captor to spend more time with the detainee. The detainee should consider dealings with his/her captors as a  battle of wits  sic: wills the detainee's will to restrict discussion to those items that relate to the detainee s treatment and return

home against the detainer s will to discuss irrelevant, if not dangerous, topics.

. As there is no reason to sign any form of document in peacetime detention, detainees will avoid signing any document or making any statement, oral or otherwise. If a detainee is forced to make a statement or sign documents, he/she must provide as little information as possible and then continue to resist to the utmost of his/her ability. If a detainee writes or signs anything, such action should be measured against how it reflects upon the U.S. and the individual as a member of the military or how it could be misused by the detainer to further the detainer s ends.

. etainees cannot earn their release by cooperation. Release will be gained by the military member doing his/her best to resist e ploitation, thereby reducing his/her value to a detainer, and thus prompting a hostile government to negotiate seriously with the U.S. Government.

. U.S. military detainees should not refuse to accept release unless doing so re uires them to compromise their honor or cause damage to the U.S. Government or its allies. ersons in charge of detained U.S. military personnel will authori e release of any personnel under almost all circumstances.

. scape attempts will be made only after careful consideration of the risk of violence, chance of success, and detrimental effects on detainees remaining behind. ailbreak in most countries is a crime  thus, escape attempts would provide the detainer with further  ustification to prolong detention by charging additional violations of its criminal or civil law and result in bodily harm or even death to the detainee.

**K. Guidance for Captivity by Terrorists:** Capture by terrorists is generally the least predictable and structured form of peacetime captivity. The captor ualifies as an international criminal. The possible forms of captivity vary from spontaneous hi acking to a carefully planned kidnapping. In such captivities, hostages play a greater role in determining their own fate since the terrorists in may instances e pect or receive no rewards for providing good treatment or releasing victims unharmed. If U.S. military personnel are uncertain whether captors are genuine terrorists or surrogates of government, they should assume that they are terrorists.

1. If assigned in or traveling through areas of known terrorist activity, U.S. military personnel should e ercise prudent antiterrorist measures to reduce their vulnerability to capture. uring the process of capture and initial internment, they should remain calm and courteous, since most casualties among hostages occur during this phase.

2. Surviving in some terrorist detentions may depend on hostages conveying a personal dignity and apparent sincerity to the captors. ostages therefore may discuss nonsubstantive topics such as sports, family, and clothing to convey to the terrorist the captive s personal dignity and human ualities. They will make every effort to avoid embarrassing the United States and the host government. The purpose of this dialogue is for the hostage to become a person in the captor s eyes, rather than a mere symbol of their ideological hatred. Such a dialogue should strengthen the hostage s determination to survive and resist. A hostage also may listen actively to the terrorist s beliefs about his/her cause however, they should never pander, praise, participate, or debate the terrorist s cause with him/her.

. U.S. military personnel held hostage by terrorists should accept release using guidance in subsection    above. U.S. military personnel must keep faith with their fellow hostages and conduct themselves accordingly.  ostages and kidnap victims who consider escape to be their only hope are authori ed to make such attempts.  ach situation will be different and the hostage must carefully weigh every aspect of a decision to attempt to escape.

# Personal Data

aw enforcement agencies need timely and accurate information to effectively work for the release of hostages. eep this data on hand, ready to give to the military security police.

**Military Member or DOD Employee          Spouse**

Full name:
 assport number:
SSN:
Rank:
 osition:
 ome address:
 hone:
 lace of birth:
 ate of birth:
Citi enship:
Race:
 eight:
 eight:
Build:
 air color:
Color eyes:
 anguages spoken:
Medical re uirements
  or problems:
Medication re uired
  and time intervals:

**Military Member or DOD Employee**      **Spouse**

rovide three signature samples:
1.
2.
.

Attach two photographs, one full length front view and one full length side view.

Attach one complete finger print card.

                     **Child 1**                  **Child 2**

Full name:
 assport number:
SSN:
Rank:
 osition:
 ome address:



 hone:
 lace of birth:
 ate of birth:
Citi enship:
Race:
 eight:
 eight:

Build:
 air color:
Color eyes:
 anguages spoken:
Medical re uirements
   or problems:
Medication
re uired and
time intervals:
 rovide three signature samples:
1.
2.

 .

## Automobiles or Recreational Vehicles

Car 1                    Car 2

Make and year:
Color:
Model:
 oors:
Style:
 icense/State:
 ehicle I :
 istinctive
   markings:

# Telephone Numbers

For additional information contact your ANTIT RR RISM  FFIC .

**Assistant Secretary of Defense**
Special  peration and  ow-Intensity Conflict
The  entagon
ashington,  .C 20 01-2 00
 0   9 -2 9 / SN: 22 -2 9

**The Joint Staff**
Attn:  4
Room 2 2 0, The  entagon
ashington,  .C. 20 1 - 000
 0   9 -  20/ SN:22 -  20

**Army**
 ead uarters  epartment of the Army
 AM -  -F
400 Army, the  entagon
ashington,  .C. 20 10
 0   9 - 491/ SN: 22 - 491

**Navy**
Chief of Naval  perations  N 4
The  entagon
ashington,  .C. 20  -  4
 0   9 -2 24/ 0  / SN: 22 -2 24

**Marine Corps**
 ead uarters  USMC
 S-1
2 Navy Anne
ashington,  .C. 20  0-1
 0    14-21 0/ SN:224-21 0

**B-38**

**Air Force**
ead uarters U.S. Air Force
Force rotection ivision
1 40 Air Force, entagon
ashington, .C. 20 0-1 40
  0   - 9 / SN: 42 - 9

**U.S. Coast Guard**
Commandant G-
2100 Second St. S
ashington, C 20 9 -0001
202 2 -0 10

# Appendix C
# Antiterrorism Individual
# Protective Measures

This appendi  contains a cut-and-fold, wallet-si e card. These safety reminders could save your life.

**Joint Chiefs of Staff**

**AT ALL TIMES**
* Vary eating establishments.
* Alternate shopping locations.
* Do not establish any sort of pattern.
* Avoid crowded areas.
* Be especially alert exiting bars, restaurants, etc.
* Know how to use the local phone system and carry "telephone change."
* Know emergency phone numbers for police, ambulance, and hospital.
* Know location of the US Embassy and other safe locations where you can find

**BOMB INCIDENTS**
* Be suspicious of objects found around the house, office or auto.
* Check mail and packages for:
  - Unusual odors.
  - Too much wrapping.
  - Bulges, bumps or odd shapes.
  - No return or unfamiliar return address.
  - Incorrect spelling or poor typing.
  - Items sent "registered" or marked "personal."
  - Protruding wires or strings.
  - Unusually light or heavy packages.
* Isolate suspect letters or packages. Do not immerse them in water. Doing so may cause them to explode.
* Clear the area immediately.
* Notify your chain of command.

**Antiterrorism Individual Protective Measures**

## FROM DOMICILE TO DUTY

* Alternate parking spaces.
* Look car when unattended.
* Look for tampering. Look under your auto. Be alert when opening door.
* Keep gas tank at least half full.
* If possible, alter routes and avoid choke points.
* Plan "escape" route as you drive.
* Watch mopeds/cyclists.
* Do not pick up hitchhikers.
* Drive with windows up and doors locked

*Remember :* **REMAIN ALERT.**

## AT AIRPORT TERMINAL

* Use concealed bag tags.
* Spend as little time as possible in airports.
* Pass through the airport security checks quickly. Once through security, proceed to a lounge or other open area away from baggage lockers. If possible, sit with your back against a wall.
* Remain alert. Be a "people watcher."

## AT HOTELS

* Do not give room number to strangers.
* Choose an inside hotel room.
* Sleep away from street side windows.
* Leave lights on when room is vacant.
* Pull curtains.
* Arrange know signals.
* Answer telephone "hello." Do not use name and rank.
* Look before you exit.
* If confronted, have a plan of action ready.
* Occasionally exit/enter through the rear entrance.
* Keep your room key in your possession at

# SECURITY

# WHILE

# TRAVELING

# Appendix D
# Antiterrorism Individual Protective Measures When Traveling

## Other Traveling Tips

- Wear inconspicuous clothing.
- Spend as little time as possible in airports, sit with your back against the wall.
- Don't give hotel room numbers to strangers.
- Leave hotel room lights on when vacant.
- Lock your hotel door.
- Keep room key in your possession at all times.
- Look for vehicle tampering.
- Drive with windows up and doors locked.

## Antiterrorism Individual Protective Measures When Traveling

- Vary your routines. Don't establish any patterns.
- Avoid crowded areas.
- Be alert when exiting bars, restaurants, etc.
- Know location of US Embassy and other safe locations.
- Know emergency numbers and how to use local phone system.
- Remain alert. Be a people watcher.

# Appendix E
# Acronyms and Abbreviations

| | |
|---|---|
| AT | antiterrorism |
| A R | area of responsibility |
| FARC | Revolutionary Armed Forces of Colombia |
| FM | Field Manual |
| FMFR | Fleet Marine Force Reference ublication |
| FM N | Faribundo Marti ara la ibercion Nacional |
| F | force protection |
| I | identification |
| | oint ublication |
| S | lysergic acid diethylamide |
| MCIA | Marine Corps Intelligence Activity |
| MC | Marine Corps rder |
| MCR | Marine Corps Reference ublication |
| mph | miles per hour |
| NBC | nuclear, biological, chemical |
| NCIS | Naval Criminal Investigative Service |
| SSN | social security number |
| UCM | Uniform Code of Military ustice |
| I | very important person |
| M | weapons of mass destruction |

# Appendix F
# References and Related Publications

## Joint Publications

Joint Publication 1-02    DOD Dictionary of Military and
Associated Terms

Joint Staff Guide 5260    Service Member's Personal Protection
Guide: A Self-Help Handbook to
Combatting Terrorism While Overseas

## Department of Defense Publications

DOD Directive
1300.7    Training and Education Measures
Necessary to Support the Code of Conduct

DOD Publication
0-2000.12-H    Protection of DOD Personnel and Activities
Against Acts of Terrorism and Political
Turbulence

## Marine Corps Reference Publication (MCRP)

3-02F/FM 21-76    Survival

## Marine Corps Order (MCO)

1510.144    Individual Training Standards (ITS) for
Antiterrorism Force Protection
(AT/FP) System

3302.1C    The Marine Corps Antiterrorism/Force
Protection (AT/FP) Program